International Studies in Educational Achievement
VOLUME 9

The IEA Study of Science II:
Science Achievement in Twenty-Three
Countries

International Studies in Educational Achievement

Other titles in the series include:

TRAVERS & WESTBURY
The IEA Study of Mathematics I: Analysis of Mathematics Curricula

ROBITAILE & GARDEN
The IEA Study of Mathematics II: Contexts and Outcomes of School Mathematics

BURSTEIN
The IEA Study of Mathematics III: Student Growth and Classroom Processes

GORMAN, PURVES & DEGENHART
The IEA Study of Written Composition I: The International Writing Tasks and Scoring Scales

ANDERSON, RYAN & SHAPIRO
The IEA Classroom Environment Study

ROSIER & KEEVES
The IEA Study of Science I: Science Education and Curricula in Twenty-Three Countries

KEEVES
The IEA Study of Science III: Changes in Science Education and Achievement: 1970 to 1984

The IEA Study of Science II: Science Achievement in Twenty-Three Countries

T. NEVILLE POSTLETHWAITE
University of Hamburg, Germany

and

DAVID E. WILEY
Northwestern University, Evanston, Illinois, USA

with the assistance of
YEOH OON CHYE
WILLIAM B. SCHMIDT
RICHARD G. WOLFE

Published for the International Association
for the Evaluation of Educational Achievement by

PERGAMON PRESS
OXFORD · NEW YORK · SEOUL · TOKYO

U.K.	Pergamon Press plc, Headington Hill Hall, Oxford OX3 0BW, England
U.S.A.	Pergamon Press Inc., 395 Saw Mill River Road, Elmsford, New York 10523, U.S.A.
KOREA	Pergamon Press Korea, KPO Box 315, Seoul 110-603, Korea
JAPAN	Pergamon Press Japan, Tsunashima Building Annex, 3-20-12 Yushima, Bunkyo-ku, Tokyo 113, Japan

Copyright © 1992 IEA

All Rights Reserved. No part of this publication may be reproduced, stored in a retrieval system or transmitted in any form or by any means: electronic, electrostatic, magnetic tape, mechanical, photo-copying, recording or otherwise, without permission in writing from the copyright holders.

First edition 1992

Library of Congress Cataloging-in-Publication Data
The IEA study of science II: science achievement in twenty-three countries / T. Neville Postlethwaite and David E. Wiley, with the assistance of Yeoh Oon Chye, William B. Schmidt, Richard G. Wolfe. -- 1st ed.
p. cm — (International studies in educational achievement: v. 9) Includes index.
1. Science--Study and teaching. 2. Science teachers.
3. Curriculum planning. 4. Academic achievement.
I. Postlethwaite, T. Neville. II. Wiley, David E.
III. International Association for the Evaluation of Educational Achievement. IV. Title: IEA study of science two. V. Series.
Q181.I44   1991   507.1--dc20   91-26614

British Library Cataloguing in Publication Data
The IEA study of science.
2. - (International studies in educational achievement; v. 9)
I. Postlethwaite, T. Neville II. Wiley, David E.
507.1
ISBN 0-08-041035-9

Printed in Great Britain by BPCC Wheatons Ltd, Exeter

To

Trudi and Annegret

Foreword

This volume is the second in a series of three presenting the results of the Second IEA Science Study.
The other two are:
Volume 1   Rosier, M. J. and Keeves, J. P. (1991) *Science Education and Curricula in Twenty-Three Countries*. Pergamon, Oxford.
Volume 3   Keeves, J. P. (1991) *Changes in Science Education and Achievement, 1970-1984*. Pergamon, Oxford.

A further report is also being prepared on *An International Assessment of Science Practical Skills*.

A First IEA Science Study was conducted in 1970 and the International Association for the Evaluation of Educational Achievement (IEA) decided in 1980 to embark on a Second IEA Science Study. The main data collection occurred in 1983/84 and in two countries in 1985.

The delay in the appearance of this volume (and the associated volumes) was caused by lack of sufficient international funds to begin the data cleaning and analyses at the appropriate time. As comparative educational research gained prominence, further funding was eventually made available and the study has now been completed. However, no report can ever present all possible analyses and it is expected that there will be many researchers throughout the world who will use the data archive for this study. Access to the data archive can be obtained upon application to:

> The Executive Director of IEA
> c/o S.V.O.
> Sweelinckplein 14
> 2517 GK THE HAGUE
> The Netherlands

It was in late 1986 that IEA called upon Neville Postlethwaite at the University of Hamburg, Germany, to raise funding for and to undertake the cleaning and analyses of the Science data. He worked closely with John Keeves, the Chairman of the International Project Council for the study and Malcolm Rosier of the Australian Council for Educational Research who was the International Coordinator.

The funding of the international part of the study was critical. Our thanks go to those agencies which financed this part of the work: the Maxwell Family Foundation in England (US$ 35,000), the Carnegie Corporation in the United States(US$ 10,000) as well as the National Science Foundation and the National Center for Educational Statistics in the United States (US$ 200,000). Needless to say the host institutions for the study - the Australian Council for Educational Research in Melbourne and the University of Hamburg in the Federal Republic of Germany bore substantial hidden costs. In each country it was either the Ministry of Education or private foundations which financed the national work.

Above all, however, our thanks go to Neville Postlethwaite of the University of Hamburg, Germany, and his co-author David Wiley of Northwestern University in the United States for undertaking the writing of this volume. They were helped in certain chapters by Yeoh Oon Chye, William B. Schmidt, and Richard G. Wolfe. IEA thanks them all.

        Tjeerd Plomp,
           Chairman of IEA.

Preface

This volume is aimed primarily at educational policy-makers at different levels of the administration in systems of education. It is also for faculty members of schools of education, their students, and for interested members of the public.

In an international empirical study of such magnitude it is impossible to analyze and report all of the data collected in one single volume. This report presents information on students, teachers, and schools for each of the populations tested (Chapter 2), the science performance of the students (Chapter 3) and student attitudes towards school and science (Chapter 4). The other chapters present analyses conducted on Population 2 data since this population is a key target group in each of the countries studied. In most countries, this is a point in school systems where nearly 100 percent of an age group is in school and before major drop-out begins. In other words, it is very near the end point of compulsory mass education in the countries examined.

The following themes are addressed: opportunities to learn science, including the content decision structure used by particular countries (Chapter 5), the teachers of science of 14-year-olds, their qualifications, their teaching load, and the various grade levels at which they taught (Chapter 6). Chapter 7 presents information on how the science curriculum is structured in the various countries and the relationship between how the curriculum is structured and student achievement. Chapter 8 examines the relationship of several home and school variables to student achievement and Chapter 9 presents brief encapsulations of science teaching and the science achievement of students in the countries involved in the study, and also summarizes the main findings of the study.

In order to keep the volume to a reasonable size, the presentation of several data analyses which had been undertaken had to be dropped. It is our intention to attempt to publish these in journal articles at a later stage. Where technical explanations have been deemed necessary, these have been presented in smaller print. We, nevertheless, hope that the non-technical reader's interest will be held.

A study of this size cannot be done by two people. Thanks are due to many persons. First, and foremost our thanks go to the 260,830 students, 22,612 teachers and 9,578 school principals who participated in the study. Secondly, to the National Research Coordinators (see Appendix A) in each of the participating research institutes who conjointly organized the research and who worked so hard on the data collection and recording. Thirdly, to the members of the International Study Committee of the Second IEA Science Study (SISS) under the guidance of Dr. Willard Jacobsen of Teachers College, Columbia University, the United States. Fourthly, to the data-processing team at the University of Hamburg: Dietmar Jungnickel, Christian Morgenstern, Andreas Schleicher, Rainer Lehmann, and Steffen Böttcher. Help was also provided by Norbert Sellin, Michael Michaels, and Frank Kühl. Fifthly, to those who have typed various drafts of the chapters: Traude Langmann, Joanne Runkel, Gunda Lemkau, Petra Lietz, Christiana Nicolitsa, Katrin Viertel, and Bettina Westphalen at the Institute of Comparative Education at the University of Hamburg. A particular debt of gratitude goes to Dieter Kotte who produced the penultimate camera-ready copy of the volume and to Jedidiah Harris who had to cope with numerous changes in order to provide the final camera-ready copy.

Three other persons were of particular help to us in the analysis and reporting stage of the work. Yeoh Oon Chye wrote Chapter 2. Richard Wolfe of the Ontario Institute for Studies in Education helped in the conceptualization of Chapters 6 and 7 and he was also responsible for the data analyses for Chapters 5, 6, and 7. William Schmidt of Michigan State University helped with Chapters 6 and 7. Chapter 2 should be cited as written by Yeoh, Chapter 5 by Wiley and Wolfe, Chapter 6 by Schmidt and Wiley, and Chapter 7 by Wiley, Schmidt, and Wolfe. All other chapters were written by us alone.

The National Research Coordinators in each country have checked the results for their countries. We are particularly grateful to Dr. John Keeves, Dr. Ken Ross and Dr. Don Spearritt for their painstaking and meticulous reading of the draft of this volume.

We hope that the results will be of use to policy-makers and that they will also inspire educational researchers throughout the world to undertake further analyses of the data that were collected in this unique comparative study.

> T. Neville Postlethwaite,
> University of Hamburg
>
> David E. Wiley,
> Northwestern University

Contents

Foreword vii

Preface ix

1. Aims, Organization and Sampling 1
2. The Contexts of Science Education 13
3. Science Achievement 49
4. The Attitude and Descriptive Measures 83
5. Opportunity and Achievement: What They Tell Us About Curriculum ... 91
6. Teachers: Teaching Loads, Grades Taught, and Subjects Taught 109
7. The Science Curriculum and Achievement 115
8. Influences on Student Achievement 125
9. Summary of Results 133

APPENDICES

A. Name and Addresses of Participating Institutions 167
B. Definitions of National Target Populations and Sampling ... 171
C. Number of Schools and Students 187
D. Test Scores for Tests 3E, 3X, and 3N (Population 3) 195
E. Test Reliabilities and Standard Sampling Errors 197
F. Mean Scores, Standard Deviations and Standard Errors of Selected Subscores ... 199
G. Report on Attitude and Descriptive Scales 201
H. Offering and Requirement Patterns With Percent Schools Covered ... 217
I. The Partial Least Squares Analysis 221

Subject Index 235

1

Aims, Organization and Sampling

A certain number of children are born in any 12 month period in a country. What do different countries do with their children in terms of education in general, and science education in particular? In some countries, all children begin school at four years of age, and in others it is at age five or six, or even seven. Some systems use age promotion and others have grade-repeating. Some have 240 days of schooling per year, others only 120 days. Some students study for six or seven hours per day and others for just a few hours. In some systems, specialist teachers of science are employed in the primary school whereas others have class teachers teaching all subjects. At the terminal grade of schooling some countries have students studying only three or four subjects whilst in other countries students may study as many as nine subjects. The average age of students in the terminal grade is 17.5 years in some systems and over 19 years in others. Moreover, the number of hours of instruction in science differs from system to system and even from student to student within systems. Some systems have science teachers who have received five or six years of pre-service training and others with only one or two years. Some systems have compulsory in-service training of teachers and in other systems it is optional. Some have well equipped laboratories and use them a lot. In other systems the reverse is true. Some curricula include more to be learned than others. These features control what students are able to learn and influence what they actually do learn. How strong are these influences? What opportunities are actually provided and what science learning results from them?

Many characteristics of schools and school systems are a result of tradition and history and some are a result of explicit policy and contemporary decisions. One factor which strongly influences the learning opportunities and experiences of students is system and school organization. Schools have differing grade spans and systems have distinctive grade span organizations. These features affect the kind of teaching expertise available and condition the delivery of curricular offerings and content opportunity in particular schools. Thus, the effects of different school organizations are reflected in student learning. This perspective raises questions about teachers and how their responsibilities are specified, and also about the courses and other activities which schools offer and require students to experience.

Do school systems really vary in the responsibilities they assign to their teachers? How are these specified and do they influence the learning experiences of students? In particular, how do teaching loads vary within and between countries? Which grade ranges are taught and how does this vary from teacher to teacher and country to country? How specialized are teachers with respect to the content taught? How and why does the degree and type of specialization vary? Does specialization relate to the grade range taught?

Other questions involve how the content focus of science courses offered and taken by students varies from country to country and school to school. What role do course requirements play in determining students' exposure to particular science subjects? Do course offerings and requirements form patterns which distinctly determine student opportunity in different countries and schools?

What characteristics of school system organization influence these patterns and how, in turn, do they affect student achievement?

The International Association for the Evaluation of Educational Achievement (IEA) undertook a study of science achievement in twenty-three countries. Data were collected from over 200,000 students, their teachers and their school principals. International tests of science achievement were constructed and administered in all systems. Data on the students' home background, aspirations, study habits and attitudes were collected. Information on the teachers' training, their teaching practices, preparation and marking as well as on their age and sex were obtained. Information on the characteristics of each school as well as the organisation of science instruction in the school was gathered.

The results can be divided into three broad categories:
- The magnitude of differences in the characteristics of students studying science in different countries and schools, the provision for education in general and science education in particular;
- The magnitudes of differences in different aspects of science achievement among countries, among schools within countries, and among students within countries;
- The factors which are related to differences among countries, among schools and among students within countries.

Chapters 2, 3, and 4 deal with the first two categories and Chapters 5 to 8 with the third category.

In early 1988, a booklet (IEA, 1988) was published presenting some preliminary results from 17 of the 23 countries, based on differences in achievement on a core test. As will be seen later, two of the target populations (see pages 3 and 4) in each system of education took a core test plus combinations of two out of four rotated tests. In the third population (the last grade of secondary schooling) all students studying science took a core test and a specialist test of biology, chemistry, or physics. In this volume, all of the test items (core plus rotated) will be taken into account and the data for all 23 countries presented.

The Organization of the Study

The project was originally proposed by IEA in 1980 and some 30 countries expressed an interest in participating. Formally, institutions from 26 school systems joined the project. These were: Australia, Canada (English), Canada (French), China, England, Finland, Ghana, Hong Kong, Hungary, Israel, Italy, Japan, Korea, Mexico, the Netherlands, Nigeria, Norway, Papua New Guinea, Philippines, Poland, Singapore, Sweden, Tanzania, Thailand, United States, and Zimbabwe. Mexico and Tanzania withdrew from the project in 1985 and 1986 respectively.

The designated research institute in a country was responsible for the funding of all data collection within the country, and the preparation and publication of a national report.

The development of all instruments, decisions about the definition of the target populations and many technical details were collective decisions. The heads of the participating centers formed an International Project Council (IPC) which was responsible for all major decisions concerning the project, ensuring that it was on timetable and within budget. Each institution appointed a National

Research Coordinator (NRC) charged with the day-to-day running of the project in the country. This person attended the meetings where all NRCs, forming the International Study Committee, met together with a Steering Committee and International Coordinator, to make all detailed decisions.

The IPC Chairman and the International Coordinator were responsible for the arrangements for all aspects of the work to be undertaken and for ensuring high quality products in terms of design, instruments, sampling, data collection and data analyses.

Appendix A presents the names and addresses of the participating institutions, the names of the IPC member and NRC for each institute, the names of the Steering Committee members, and the names of persons who worked at the International Coordinating Center and the Data Processing Centers for periods of more than six months.

Target Populations and Sampling

Rosier (1987, p. 106) has maintained that "science is included in the school curriculum for two reasons. First, it has a role in helping students to understand their environment and to develop skills in the application of scientific methods to the solution of problems. Secondly, it provides basic training for those students who will subsequently follow careers in science and technology".

This understanding of science in the school curriculum was reflected in the way in which populations were selected for testing. There had also to be comparability with the 1970 study.

Target Populations

Three levels in a school system were selected:
1. A point somewhere towards the end of what is normally called primary schooling or the first cycle of education, where students are usually taught by a classroom teacher and not a science specialist
2. A point somewhere toward the end of full-time compulsory education where in most countries 100 percent of an age group is still in school
3. The final year of full-time secondary schooling.

The first two populations, in general, represent the assessment of science education for all. The third population represents an assessment of science achievement for all students who reach the final grade of secondary school. All students studying biology, chemistry and physics formed one part of the third population and those not studying science another part.

The internationally agreed-upon operational definitions of these target populations became:

Population 1: All students aged 10:0 to 10.11 on the specified date of testing OR all students in the grade where most 10 year olds were to be found on the specified date of testing.

Population 2: All students aged 14:0 to 14:11 on the specified date of testing OR all students in the grade where most 14 year olds were to be found on the specified date of testing.

The first option in these definitions corresponded to the definitions used for the First IEA Science Study (Comber and Keeves, 1973). The second option was included for the practical reason that it was harder to obtain age samples of students in most countries.

Once the tests had been prepared, some developing countries realized that the content of the tests was too difficult for the IEA defined grade groups in their countries, and requested permission to be able to test grade groups for which the content was appropriate. This permission was granted although it was realized that there could be some problems with direct international comparisons. On the other hand, by testing at a higher, more appropriate grade level, it was likely that more valid measures of science achievement would be obtained. It was very important to have achievement criteria that were as valid as possible in order to facilitate examination of the relationships between explanatory factors and student science achievement. In addition, some developed countries chose grades in which 14-year-olds were enrolled but in which 15-year-old or older students were also enrolled. In these countries the average age of enrolled students also exceeded the planned range. Consequently, in some countries the above definitions were extended to include one or two grades later in the system than that of the originally agreed population.

Population 3: All students in the final year of full-time secondary education. There were two important subgroups of this population. The first group comprised those students (Population 3 - Science) studying science at a level which would be high enough to enable them to proceed to tertiary studies in science. These students were divided into several sub-populations as were appropriate in the different countries.

- 3E - all students studying earth science (not reported in this volume);
- 3B - all students studying biology;
- 3C - all students studying chemistry;
- 3P - all students studying physics;
- 3X - those students studying a general science course (not reported in this volume).

In practice, very few countries tested their students who were taking an earth science course (Sub-population 3E) or a general science course (Sub-population 3X).

The second group comprised those students not studying science at a level which would enable them to proceed to tertiary studies in science. These students formed a second population defined as:

- 3N - all students at the final year of full time secondary education not currently studying science subjects which would enable them to continue the study of science at a university or institute of higher education.

It was agreed that the testing would take place, where possible, within the last three months of the school year.

Table 1.1 presents the age of entry to school, the grades tested, the percentage of an age group in school, and the mean and standard deviation of age for Populations 1 and 2 in each participating country. Table 1.2 presents similar data for Population 3.

It should be noted that few countries selected the first option, of an age-defined population: Australia for Populations 1 and 2, and Ghana for Population 2. Canada (English) included English-speaking students from nine provinces. The provinces omitted were Québec and the Northwest Territories. Canada (French) included students in francophone schools where science was taught in French in the province of Québec and in all of the other provinces for Population 1, six other provinces for Population 2, and five other provinces for Population 3. China obtained random samples in Beijing, Tianjin, and Taiyuan, the provincial capital of Shan Xi. Nigeria included 60 percent of its 19 provinces in the target population. In Australia, Israel and the United States, because of low levels of response in 1983 at the Population 3 level, it was considered necessary to increase sample sizes by returning a year later to the schools in 1984 which were drawn in the original sample but which did not participate. The United States undertook two sets of data collection. The first testing (Phase 1) took place in 1983/84. However, the United States undertook a second testing in 1986.

TABLE 1.1   Percentage of an age group in school and mean ages of Populations 1 and 2

Country	Age of Entry	Population 1				Population 2			
		Grade Tested	%in School	Age[a] Mean	sd	Grade Tested	%in School	Age[a] Mean	sd
Australia	6	4,5,6	99	10:6	3.3	8,9,10	98	14:5	3.3
Canada (Eng.)	6	5	99	11:1	7.1	9	99	15:0	6.1
Canada (Fr.)	6	5	99	11:1	6.4	9	82(94)[b]	15:3	8.4
China[c]	6	-	-	-	-	9	37	15:8	8.7
England	5	5	99	10:3	3.6	9	98	14:2	3.6
Finland	7	4	99	10:10	4.1	8	99	14:10	4.1
Ghana	6	-	-	-	-	9	6(43)[d]	16:1	20.2
Hong Kong	6	4	99	10:5	9.8	8	99	14:7	10.9
Hungary	6	4	99	10:3	5.2	8	92	14:3	4.7
Israel	6	5	99	10:9	3.4	9	99	14:9	3.4
Italy (Grade 8)	6	5	99	10:9	5.2	8	99	13:11	8.6
Italy (Grade 9)	-	-	-	-	-	9	72	14:8	3.2
Japan	6	5	99	10:7	3.5	9	99	14:7	3.5
Korea	6	5	99	11:2	7.4	9	99	15:0	7.2
Netherlands	6	-	-	-	-	9	99	15:6	12.5
Nigeria	6	6	92	12:1	6.5	10	na	16:2	6.8
Norway	7	4	99	10:11	4.0	9	99	15:10	4.0
Papua New Guinea	7	-	-	-	-	10	11	17:1	n.a.
Philippines	7	5	97	11:1	11.3	9	60	16:1	18.9
Poland	7	4	99	10:11	5.4	8	91	15:0	5.8
Singapore	6	5	99	10:10	4.9	9	91	15:3	9.2
Sweden (Grade 7)	7	3	99	9:10	3.7	7	99	13:10	4.8
Sweden (Grade 8)	7	4	99	10:10	4.1	8	99	14:10	3.8
Thailand	6	-	-	-	-	9	32	15:4	8.9
U.S.A. (Phase 2)	6	5	99	11:3	6.9	9	99	15:3	9.1
Zimbabwe	6,7,8	-	-	-	-	9	30	16:1	17.7

-   this population not tested.
- a   the mean age is presented in years and months and the standard deviation in months.
- b   94% of the age group is in school. Only 82% was tested because the vocational stream was omitted.
- c   China tested only in Beijing, Tianjin, and Taiyuan (the provincial capital of Shan Xi). The estimate of the % in school relates to these three provinces only. For China as a whole the estimate was 31% in 1985.
- d   43% of an age group is in school but since middle school students were not tested, those tested represent only 6% of an age group.
- na   data not available.

TABLE 1.2 Percentage of an age group in school and mean ages in Population 3

Country	Grade Tested	All students in final grade			Biology		Chemistry		Physics		Non-Science		Average No. of Subjects Studied
		% in School	Age[a] Mean	sd	% in School	Age[a] Mean	% in School	Age[a] Mean	% in School	Age[a] Mean	% in School	Age[a] Mean	
Australia	12	39	17:3	11	18	17:1	12	17:3	11	17:3	10	17:6	5
Canada (Eng.)	12/13[b]	68	18:3	11	28	18:2	25	18:4	18	18:4	-	-	6
Canada (Fr.)	11/12[c]	67(79)[d]	17:2	09	7	17:2	37	17:1	35	17:1	12	17:3	6
England	13	20	18:0	06	4	18:0	5	18:0	6	18:0	10	18:0	3
Finland	12	41(63)[de]	18:6	07	41	18:7	16	18:6	14	18:7	f	-	9+
Ghana	13	1.2	18:8	18	0.2	18:8	0.6	18:10	0.6	18:11	-	-	3
Hong Kong (F.6)	12	27	18:3	13	12	18:5	20	18:4	20	18:4	-	-	6
Hong Kong (F.7)	13	20	19:2	11	7	19:2	12	19:3	12	19:3	-	-	5
Hungary	12	18(40)[dg]	18:0	04	3	18:0	1	18:1	4	18:0	9	18:1	9+
Israel	12	65	17:6	09	20	17:7	8	17:7	12	17:7	25	17:7	7
Italy	12/13	34[h]	19:0	13	4	19:5	1	19:2	13	19:2	10	19:1	7
Japan	12	63(89)[d]	18:2	04	12	18:1	16	18:2	11	18:2	35	18:2	7+
Korea	12	38(83)[d]	17:9	08	38	17:11	37	17:10	14	17:11	-	-	9+
Norway	12	40	18:9	07	4	18:11	6	18:11	10	18:11	24[i]	18:11	7
Papua New Guinea	12	1.1	18:7	10	-	-	-	-	-	-	0.4	18:7	3
Poland	12	28	18:6	05	9	18:7	9	18:7	9	18:7	8	18:3	9+
Singapore	12/13	17	18:1	08	3	18:0	5	18:0	7	18:0	8	18:3	6
Sweden(j)	12	28	19:0	11	5	18:11	13(6)[j]	19:0	13	19:0	15	19:0	9+
Thailand	12	14(29)[d]	18:3	08	7	18:3	7	18:3	7	18:2	7	18:4	6
U.S.A. (Phase 2)	12	83	17:7	09	12	17:5	2	17:8	1	17:10	-	-	5

* This population not tested.
[a] The mean age is given in years and months. The standard deviation is presented for the All group only and is in months.
[b] In Ontario, it was Grade 13.
[c] The final grade in school in Quebec is 11 and in other provinces 12. The age calculations are based on the total sample of francophone schools in all provinces. The percent of an age group in any one subject group is based on Quebec figures only.
[d] The difference between the percent given and the percent in parenthesis is accounted for by the vocational students.
[e] 63 percent of an age group is in school at this level. 18 percent are in vocational schools and four percent in evening school. The 41 percent are in full time day schooling.
  In Finland all students in academic schools take biology, consequently there were no non-science students in the academic schools, which account for 41 percent of the age group.
[f] The 18 percent of the age group is those in academic secondary schools studying science.
[g] This figure is exact. The figures for the percentages of an age group in school in Italy for the specialist populations are estimates.
[h] Vocational school students are excluded from this figure.
[i] In 1984 in Sweden, 90 percent of an age group was enrolled in upper secondary education (Grades 10-12) and 80 percent of an age group complete upper secondary education (Grade 11). Thirteen percent were enrolled in science tracks and 15 percent in non-science tracks in Grade 12. The remainder took a two year vocational or general track and left school after Grade 11. Six percent of students of the age group in the technological track continued at school for a thirteenth year. These students were not tested. Although only six percent of an age group study chemistry, 13 percent of the age group was tested.

Aims, Organization and Sampling

The purpose of the second study in the United States (Phase 2) was to assess process skills (which had not been included in the first round of testing) and to administer the core test for correlational purposes with process skills. At the same time, the core test could be examined against the 1983/84 baseline data. At Population 3 level, the first round of testing included both first and second year physics students in Grade 12 and Population 3N, the non-science students, only. In the second round of testing, second year biology, chemistry and physics students were tested. Some were in Grade 10, some in Grade 11 and some in Grade 12 (see Appendix B). This problem was addressed in the first meeting of the advisory panel to the second U.S. National Science Foundation-funded portion of the study. This meeting took place at Teachers College, Columbia University in New York after data processing had progressed considerably. In that meeting, it was directed that only data from the second round of testing be used as the student data from the first testing round appeared to have sampling inadequacies which were difficult to evaluate and which possibly biassed estimates of sample means and proportions. This "direction" was followed but it created two problems: the first was that multivariate analyses for the United States became virtually impossible because many variables were not administered in the second round of testing; the second was that several items were dropped at the second round of testing, since no rotated tests were employed at Populations 1 and 2, and several items were dropped from the biology, chemistry and physics tests. Hence, the "direction" could not be followed for all data analyses as it would have eliminated comparisons involving the United States. In the cases where these data were used, the purpose was not to estimate population means or proportions, but to explore variability and assess relationships of particular student and school characteristics to achievement.

Care must be taken when comparing the mean scores for science achievement in Populations 2 and 3. It will be seen for Population 2 that the grade level selected for testing does not always accord with the modal grade of 14-year-olds. Furthermore, the proportion of an age group in school at that level - although nearly 100 percent in all developed countries - varies a great deal among the developing countries. At the Population 3 level, there are many differences among countries which should be borne in mind when examining science achievement differences: whether the last grade is Grade 11, 12 or 13; the definition of those taking science and those not taking science; the number of subjects studied; the associated issue of the cumulative time devoted to studying science; the proportion of an age group studying the subject in question and the average age of the sample. Part 1 of Appendix B presents detailed descriptions of each of the target populations in each country.

Sampling

A detailed report on the sampling procedures used can be found in 'Administration and Sampling Report of the Second International Science Study' (Rosier, 1989). This report is filed with ERIC. What follows is a very brief overview.

Samples were either three or two stage probability samples. Two stage samples involved a sample of schools (or, in some cases, classes) with either students or an intact class being drawn at random from within the selected schools. The samples were typically designed to yield 2.5 percent or less of a student standard deviation for one standard error of sampling of a mean value. All sampling plans had to be approved by a sampling referee (M.J. Rosier).

It should be noted that the use of intact classes in all countries, except Australia, Canada (French), England, Ghana, Italy, and Poland, increased the magnitudes of the sampling errors, and some countries, namely Japan, Canada (English) and the Philippines increased the sizes of their designed samples to compensate for this. However, it is important to note that countries probably achieved higher response rates by using intact classes than if they had used selected students from several classes. This is because schools are more willing to participate in a research study if intact classes are used, causing less disruption for the school than having to collect students together from several classes.

Whereas the students are a probability sample of all students in the defined target population, the teachers are not a probability sample of all science teachers. They are either the class teachers or the teachers teaching science at the defined grade level in the schools. At Population 1 level only Hong Kong, Israel, Norway, the Philippines, Singapore and the United States have taken the class teacher. In the other countries, all teachers teaching science at that grade level were taken. In Population 2 level the countries where one class science teacher only received a questionnaire were Hong Kong, Israel, Norway (Grade 9), the Philippines, Thailand and the United States. In other cases, it was either all teachers teaching science at that level in the school or the different teachers of the various branches of science. At Population 3 level, it was the teachers of the various branches of science who received questionnaires. For descriptive statistics about teachers all teachers for whom data were returned were used. In some cases, this included all teachers teaching science at a particular grade level. In other cases, it included only those teachers teaching the students tested. It should be noted, that in Chapter 6 it is the teachers who were teaching the students tested who were used in the analyses.

Part 2 of Appendix B presents a brief overview of the sampling procedures used for each of the populations in each country.

After the testing program had been completed the data from the executed sample were recorded on file. The data were then cleaned and an achieved sample obtained. The cleaning rules for constituting the achieved sample were:

Student	A student would only enter the cleaned file if he/she had attempted at least three items on the core test.
Teacher	A teacher would only enter the cleaned file if he/she had answered at least one question on the teacher questionnaire.
School	A school would only enter the cleaned file if at least three students had entered the student cleaned file. Thus, a school might be entered on the cleaned school file as an I.D. number even though no school questionnaire had been completed.

Aims, Organization and Sampling

Test scores
A core test score would be computed for a student if at least three items had been attempted. A total test score would be computed if a student had attempted at least three items on the core test and one item on the rotated tests.

Weights
Stratum weights would be calculated. The number of students in the achieved sample would be the number arrived at after the above cleaning process.

Independent variables
A variable would be dropped if more than 20 percent of respondents, whether students, teachers, or schools, on the cleaned file had not provided information.

Response Rates, Weighting, Standard Errors of Sampling and Marker Variables

Response Rates. The number of schools and students, after the above cleaning process, formed the achieved samples. The tables in Appendix C present the response rates for the different populations for schools and students. In each table, the numbers of schools/students in that target population is presented, the designed sample, then the executed sample (the executed sample is the schools and students supplying data), and finally the achieved sample (i.e., after cleaning). Appendix C1 presents the numbers of schools and students in Population 1, C2 for Population 2, C3 for all students in Population 3, C4 for Population 3 biology students, C5 for the chemistry students, C6 for the physics students, and C7 for the students not studying science (3N). The response rates for both schools and students are presented in the form:

$$\frac{n \text{ in achieved sample}}{n \text{ in designed sample}} \times 100$$

At Population 3 level, the total number of students in Population 3 were known but only in very few cases were the target population numbers known separately for those studying biology, chemistry and physics. From Table 1.2 above it is clear that there was overlap among the students studying the various science subjects. In most countries, a probability sample of schools was drawn and then sub-samples drawn of those studying the various subjects. Where students studied more than one science subject they were typically allocated at random to only one of the science tests.

Where target population figures existed for the separate subjects, these were used in the calculation of weights. Where such figures were not known, the total Population 3 figures were used. Which figures were available can be seen in Appendices C1 to C7.

Weighting. The percentage of loss from the designed to the achieved sample varied from stratum to stratum. To correct for this differential loss when calculating national scores, stratum weights were calculated and applied.

For nearly all strata, the stratum weights ranged from 0.8 to 1.4. About seven percent exceeded 2.0 and about two percent fell below 0.4. The more extreme values were all related to small strata in countries where the number of students was large.

These stratum weights were used not only for test scores but also for the production of student univariates. Teacher and school univariates were unweighted.

It should be pointed out that all Israeli data were unweighted since no population figures were made available. At Population 3 level no weighting was necessary for Papua New Guinea and Singapore since all students in the target population were tested.

Sampling Errors and Sample Bias. In order to have estimates of the 'goodness' of sampling two steps were taken.

First of all, the standard errors of the national test means for Populations 1 and 2, as well as for Population 3 biology, chemistry and physics and for Population 3N are presented in Appendix C.8. The standard errors were calculated by jackknife repeated replication by dropping one primary sampling unit (school or class) at a time (see Tukey, 1977).

Secondly, a comparison was made between a target population statistic and a sample statistic on one variable in order to check for sample bias. The variable was percent male students in the target population. The population values which were reported by nearly all countries were the percentage of male students in the grade tested. The population percentages, sample percentages and standard errors of the sample percentages are reported in Appendix C.8. The standard errors of percent male were calculated in the same way as the standard errors of sampling of mean scores by dropping a school at a time. As a rule of thumb, if there is a difference between the population and sample statistics of more than five percent, or two standard errors, this raises a warning flag about possible bias in the sample. For Populations 1 and 2 it will be seen that, with the exception of the Philippines, the sample estimates are close to the population values. Care needs to be taken with the Population 3 figures. For Population 3, there is the problem of the National Centers not having access to the population statistics of the number of males and females studying biology, chemistry, and physics separately or in some combination at that level. Most ministries of education do not record these figures. In Australia they were known but for other countries where different percentage values are given for each subject they are 'guesstimates'. Where the same figure appears for all three subject areas the overall percentage male value is given at that grade level. There may be a bias in the sample for Finland in biology (but possibly not in chemistry and physics) and in Italy and the United States. The United States sample may not be biassed since it is possible that the population statistics are inaccurate.

Data Collected

As stated earlier, the main aims of the study were: to describe science education in the participating countries; to examine between-country achievement differences and, where possible, explain them; and, to attempt to explain differences in achievement between students within countries.

Data were, therefore, collected on selected structural features of each educational system, on the students and their home and educational backgrounds, and on what happens in schools (and, in particular, what science teachers do in teaching science). Special attention was paid to the intended curriculum in science (the syllabus), the implemented curriculum (what the teachers actually teach), and the achieved curriculum as measured through the science achievement tests and the science attitude scales which were administered. More details are given on these measures in Chapters 3 and 4. The figures below for the number of schools and students tested are given to indicate the magnitude of the study.

	Schools	Teachers	Students
Population 1	3,096	5,065	81,855
Population 2	3,658	9,830	94,972
Population 3	2,828	7,860	85,449
Total	9,582	22,755	262,276

It was well known that not all data would be used in reporting the study but, since it was envisaged that many scholars would wish to use the data bank for the purposes of secondary analysis, it may be useful for them to know that data exist on many variables that are not reported in this volume.

In all, the Second IEA Science Study had approximately 1,500 megabytes of information. Many of the files handed in by the National Centers contained numerous errors and inconsistencies and the cleaning of the files was a lengthy and tedious process. The analyses consisted of item and scale analyses, the computation of weighted univariate statistics on all background variables, and the undertaking of many multivariate analyses. This was a considerable undertaking with only six part-time personnel working to a restricted financial and time budget.

Chapter 2 will review brief descriptions of selected student, teacher and school variables across the three different populations. One purpose is to outline the varied contexts of science education across the different countries in 1984/85. In order to interpret the data presented in Chapter 2 it is important for the reader to be able to refer to the information which has been given in this chapter.

References

Comber, L.C. and Keeves, J.P. 1973 *Science Education in Nineteen Countries*. Almqvist and Wiksell, Stockholm.
IEA 1988 *Science Achievement in Seventeen Countries*. Pergamon, Oxford.
Rosier, M.J. 1987 The Second International Science Study. In: *Comparative Education Review*, 31(2):106-128.
Rosier, M.J. 1989 *Administration and Sampling Report of the Second International Science Study*. ERIC.
Tukey, J.W. 1977 *Exploratory Data Analysis*. Addison-Wesley, Reading, Mass..

2

The Contexts of Science Education[1]

As was stated at the beginning of Chapter 1, countries organize their systems for educating their children in different ways, dependent upon the different economic and social conditions as well as on different philosophies of education. The contexts in which the children study in the various countries are different.

The aim of Chapter 2 is to present a comparison of the contexts of science education across the 24 different systems of education. Some countries collected data for two grade groups for a particular population. These were Sweden with Grades 3 and 4 at Population 1 level, and Grades 7 and 8 at Population 2 level, Italy with Grades 8 and 9 at Population 2 level and Hong Kong with Grades 12 and 13 (Forms 6 and 7 of secondary school) at Population 3 level. All are reported.

The chapter begins with the overview of a few selected characteristics of each participating country. The data in turn are tabulated against the background of science education, seen either in terms of the nation's per capita GNP or, differences in the level of science achievement. For example, Table 2.1 offers the reader the opportunity to examine the variables not only within a system but also across the nations of varying levels of per capita GNP.

Following this, the chapter then examines the characteristics of the students, teachers and schools separately for Populations 1 [modal age 10+], 2 [modal age 14+] and 3 [the last year of secondary school]. The patterns of variation are examined in Tables 2.2 to 2.11, in relation to the corresponding levels of school science achievement. For ease of reading and comparative analysis, the variations within and between the different systems are sequenced in order of their respective levels of school science achievement.

In some cases, certain national data were sent in by National Centers. This is the case for all data in Table 2.1. and for the variable 'Hours per year total instruction' presented in Tables 2.4, 2.8, and 2.11 describing characteristics of schools in the sample.

In some tables, readers will see figures surrounded by square brackets []. These were not on the data files sent in by each system but were supplied by the National Centers from other data sources.

All other data come from the SISS data files. It will be recalled that the samples consisted of schools drawn with a probability proportional to size. Therefore, the figures in the tables represent the characteristics of students, teachers, and schools from the point of view of a typical student in the sample drawn. It must be borne in mind that the data were collected by questionnaires and there are limits to this mode of data collection. Responses, particularly by younger children, may not be fully accurate and the data collected may provide only an incomplete or indirect indication of the phenomena in which one is interested (see Thorndike, 1973, pp. 42-43). In some countries, information

---

[1] This chapter has been written by Yeoh Oon Chye. Dr. Yeoh is Head, School of Education and Development Studies, Institute of Education. He was the National Research Coordinator for this study for Singapore.

TABLE 2.1 Selected national characteristics of the participating countries 1984/85 in relation to the World Bank classification of GNP/capita

World Bank Classification [GNP/capita]		Group 4 US$ 4,301 or more											
Countries		ISR	HKO	ITA	SIN	ENG	NET	JAP	FIN	AUS	SWE	CAN	NOR	USA
National Characteristics GNP/cap.[1984/85]		5,090	6,370	6,430	7,470	8,590	9,540	10,650	10,800	11,060	11,880	13,310	13,970	15,540
Size of Population [in 000]		4,209	5,364	56,983	2,531	56,383	14,420	120,018	4,882	15,544	8,337	25,124	4,140	237,019
Land Area [km^2] [in 000]		20	1	na	0.6	244	42	378	337	7,700	450	9,976	324	9,404
Education Budget [% state budget]		9.2	18.7	9.8	10.8	11.5	16.8	18.1	12.5	13.2	12.2	15.2	12.8	17.1(d)
Years Compulsory Schooling		11	9	8	10(a)	11	11	9	9	9	9	10	9	10
Days Schooling per Year	Pop 1	200	207	na	186	190	195	210	190	180	178	180	185	180
	Pop 2	190	181	na	180	190	195	210	190	180	178	180	185	180
	Pop 3	170	181	na	180	190	195	210	na	160	178	195	185	180
Last Grade General Teaching		6	6	na	3	6	na	6	6	6-7	6	6	6	6
% of an Age Group in School	Pop 1	99	99	99	99	99	-	99	99	99	99	99	99	99
	Pop 2	99	99	99(h) 72(k)	91	98	99	99	99	98	99	82(b) 94(b)	99	99
	Pop 3	65	27(f) 20(f)	34	17	20	-	63 89(e)	41 63(e)	39	28	68 79(e)	40	83
Average No. of Subjects at Final Grade	Pop 3	7	6(f) 5(f)	7	6	3	-	7+	9+	5	9+	6(b) 6(b)	7	5
Mean Students' Age [Yr:Mth]	Pop 1	10:9	10:5	10:5	10:10	10:3	-	10:7	10:10	10:6	9:10(j) 10:10(i)	11:1(b) na	10:11	11:3
	Pop 2	14:9	14:7	13:11(h) 14:8(h)	15:3	14:2	15:6	14:7	14:10	14:5	13:10(j) 14:10(j)	15:0(b) 15:3(b)	15:10	15:3
	Pop 3	17:7	18:4(f) 19:2(f)	19:0	18:1	18:0	-	18:2	18:7	17:3	19.0	18.3(b) 17.2(b)	18.11	17:7
Pupil/Teacher Ratio	Pop 1	na	30.2	13.7	25.6	25.0	na	29.3	18.5	23.5	-	18.5(b) 21.0(b)	na	18.4
Teacher Salary [% GNP/capita]	Pop 1	na	169	171	134	184	na	193	199	207	122	na 202(b)	122	152
Science Achievement [in % score]	Pop 1	41.5(k)	50.9	59.2	51.8	53.8	na	66.4	66.0	56.9	55.9(k) 62.8(k)	61.7(b) 61.4(b)	55.5	54.8(l)
	Pop 2	58.5(k)	55.0	52.4(h) 59.8(h)	56.4	55.9	63.7	66.8	60.3	58.8	56.0(l) 60.3(l)	61.6(b) 58.5(b)	59.3	54.8(l)
	Pop 3^m	C1	A4 A1	D	A3	A2	na	B2	C2	D	C2	D	B4	D

(a) Schooling is not compulsory. However, most pupils complete 10 years of schooling.
(b) Canada: English and French respectively.
(c) Based on Unesco Yearbook 1973.
(d) Based on Unesco Yearbook 1981.
(e) Vocational and academic students included.
(f) Hong Kong (Forms 6 and 7) in Population 3.
(g) In Poland, Biology is taught separately in Grade 4.
(h) ITA8 and ITA9 in Population 2.
(i) SWE3 and SWE4 in Population 1.
(j) SWE7 and SWE8 in Population 2.
(k) Unweighted percentage mean scores in Israel only.
(l) Based on the core items only of Pop 1 and Pop 2. The test items in the Rotated Forms A/B/C/D are excluded.
(m) Achievement Group A = At least 55% score or better in B/P/C.
Achievement Group B = At least 50-60% score in B/P/C.
Achievement Group C = At least 36%-49% score in B/P/C.
Achievement Group D = Consistently less than 50% in B/P/C.

A1-A4/B1-B4 = Rank order in Group A and B achievement.

- not participating
na not available

The Contexts of Science Education

TABLE 2.1 (ctd.) Selected national characteristics of the participating countries 1984/85 in relation to the World Bank classification of GNP/capita

World Bank Classification [GNP/capita]		Group 3 US$ 1,636 - US$ 4,300			Group 2 US$ 401 - US$ 1,636					Group 1 US$ 400 or less	
Countries		HUN	POL	KOR	PHI	PNG	NIG	ZIM	THA	CHN	GHA
National Characteristics GNP/cap.[1984/85]		2,060	2,110	2,150	660	700	740	760	840	310	370
Size of Population [in 000]		10,668	36,914	40,127	53,380	3,422	96,485	8,099	50,023	1,029,156	12,309
Land Area [km^2] [in 000]		93	313	na	1,295	463	942	na	na	9,600	238
Education Budget [% state budget]		6.4	11.4	24.2	7.0	16.0	11.6	16.0	21.1(c)	8.1	21.5
Years Compulsory Schooling		8	8	6	6	0	na	7	6	9	9
Days Schooling per Year	Pop 1	180	200	220	194	207	na	190	200	192	169
	Pop 2	180	200	220	194	207	na	190	200	192	158
	Pop 3	160	155	220	-	207	na	190	200	192	158
Last Grade General Teaching		4-5	3(g)	6	4	6	na	7	6	2	na
% of an Age Group in School	Pop 1	99	99	99	97	-	na	-	-	-	-
	Pop 2	92	91	99	60	11	na	30	32	37	43
	Pop 3	18 40(e)	28	38 83(e)	-	1.1	-	-	14 29(e)	-	1.2
Average No. of Subjects at Final Grade	Pop 3	9+	9	9+	-	3	-	-	6	-	3
Mean Students' Age [Yr:Mth]	Pop 1	10:3	10:11	11:2	11:1	-	12:1	-	-	-	-
	Pop 2	14:3	15:0	15:0	16:1	17:1	16:2	16:1	15:4	15:8	16:1
	Pop 3	18:0	18:7	17:11	-	18:7	-	-	18:3	-	18:10
Pupil/Teacher Ratio	Pop 1	11.8	19.1	51.8	32.9	na	28.6	na	na	na	na
Teacher Salary [% GNP/capita]	Pop 1	79	94	212	190	na	na	na	na	na	na
Science Achievement [in % score]	Pop 1	61.7	52.5	65.7	42.3	na	35.1	na	na	na	na
	Pop 2	70.7	59.5	61.0	39.7	55.3	42.2	42.8	56.7	60.0	46.7
	Pop 3^m	B1	B3	D	-	-	-	-	D	-	B1

(a) Schooling is not compulsory. However, most pupils complete 10 years of schooling.
(b) Canada: English and French respectively.
(c) Based on Unesco Yearbook 1973.
(d) Based on Unesco Yearbook 1981.
(e) Vocational and academic students included.
(f) Hong Kong (Forms 6 and 7) in Population 3.
(g) In Poland, Biology is taught separately in Grade 4.
(h) ITA8 and ITA9 in Population 2.
(i) SWE3 and SWE4 in Population 1.
(j) SWE7 and SWE8 in Population 2.
(k) Unweighted percentage mean scores in Israel only.
(l) Based on the core items only of Pop 1 and Pop 2. The test items in the Rotated Forms A/B/C/D are excluded.
(m) Achievement Group A = At least 55% score or better in B/P/C.
Achievement Group B = At least 50-60% score in B/P/C.
Achievement Group C = At least 36%-49% score in B/P/C.
Achievement Group D = Consistently less than 50% in B/P/C.

A1-A4/B1-B4 = Rank order in Group A and B achievement.

- not participating
na not available

relating to the home was collected from the parents but in most countries reliance was placed on students' responses.

National Context Variables

Table 2.1 presents certain national context variables.

The countries vary in population size and land area. The density of population per square kilometer also varies. Hong Kong and Singapore have relatively dense populations with 5,364 and 4,218 people per square km. From the available data, Australia appears to be least dense [2.02/km^2] but it must be recognized that two-thirds of Australia's population lives in New South Wales and Victoria. The GNP per capita varies dramatically from US$ 310 in China to US$ 15,540 in the United States. These per capita GNP figures have been categorized into one of four categories used by the World Bank. In 1984, countries varied in the percentage of the state budget which was allocated to education. If the percentage of the state budget allocated to education is viewed as an indicator of the priority that countries award to education, then Korea, Thailand and Ghana accord high priority to education whereas Hungary and the Philippines would appear to place low financial emphasis on education. At a glance, it is evident that Ghana and Papua New Guinea give very high priority to education in spite of their low per capita GNP. Certainly, among the Group 4 nations, the expenditure on education varies widely, ranging from 9.2 percent for Israel to 18.1 percent or more of the State Budget by Hong Kong and Japan. Notwithstanding variation in circumstances, a measure of educational productivity [say school science achievement] is not necessarily linked directly to the extent of expenditure on education.

In Table 2.1 the years of full-time compulsory schooling range from zero in Papua New Guinea to 11 in England, Israel and the Netherlands. Schooling in Singapore is not compulsory. However, most students complete at least eight years of schooling. The number of days of actual schooling per year varies from 155 in Poland to 220 in Korea. Such figures tell nothing about how many hours per day are involved but this will be taken up later in this chapter. Where the number of days per year in Population 3 is lower than Population 2, this is usually because the students are given time off from school in order to study for their final examinations.

Population 1

Students

Table 2.2 presents selected characteristics of the students in Population 1. At this level in the school system all countries have about 50 percent **boys** and 50 percent **girls** enrolled in school except for the Philippines where about five percent of boys have already dropped out of school. The average **number of children** in the family ranges from 2.2 in Norway to four or more in Nigeria and the Philippines. The **number of books** in the home ranges from relatively few books in Israel (14) and under 100 in Korea, the Philippines, Italy, Poland and Singapore to over 200 in Australia, England, Norway and Sweden. It must be noted that the actual figures should be treated with caution. The figures result

The Contexts of Science Education 17

from recoding a six category response. But, the relative magnitude provides a measure of the differences between countries in books available at home. The number of **years of education of the parents** ranges from four to 12 years. Care must be exercised when dealing with the estimates of father's and mother's education throughout the volume. The variable in its original form was truncated in some countries because the question in the student questionnaire had as the lowest category '9 years or fewer'. This was clearly inappropriate for many countries and in some developing countries, very inappropriate. Some countries included categories of fewer years of education but others did not. Some countries omitted or modified the question. Thus, readers are asked to bear this in mind when examining the figures.

The **size of the school classes** in which these students find themselves are mostly 20 to 30 students per class but there are high figures for Hong Kong (38), Japan (38), Korea (55), the Philippines (40) and Singapore (37). These are all countries with high population densities in the East and Southeast Asian Region. The minor discrepancies between Tables 2.2 and 2.3 for this variable are because Table 2.2 is based on replies from students in the sample and Table 2.3 from the teachers in the schools.

It is clear that the children in the developing countries tend to come from larger **families**, with few **books** in the home and with **parents** having less education than those in other countries. At the same time, there is quite a large variance within each country on each of these indicators.

Teachers

Table 2.3 presents data on selected teacher variables.

The teachers teaching Population 1 science tend to be **female**. Japan has 50 percent while Korea (58.3%) and Nigeria (74.1%) have more males than females teaching in the primary grades. For the majority of countries the **average age of teachers** is in the late 30s but Sweden's teachers are in their late 40s and Nigeria in their late 20s. The number of **years post-secondary education** received varies considerably from one year in Italy and Nigeria to four or more years in Canada (French) and Finland. Although the Philippines has 4.5 years of higher education it must be remembered that the Philippines has only 10 years of primary and secondary schooling and hence if they had 12 years of secondary school, the equivalent would be 2.5 years of post-secondary education. The percentage of the **post-secondary education devoted to science** varies enormously from three to four percent in Philippines and Italy to over 40 percent in Israel and Poland. Hungary, Israel, the Philippines and Poland also have greater amounts of in-service training ranging from three to four days per year. There is a clear linear relationship between the **average age of teachers** and the corresponding number of **years of teaching experience**.

Finland, Italy, Canada (Fr.), the Philippines and the United States teachers are in the 40s with 15-20 years of **teaching experience**. In contrast, the teachers in six countries (Israel, Hungary, Singapore, Poland, Hong Kong and Japan) have also acquired 15-20 years of experience but they fall in the age range of 30-40 years only. This implies that primary school teachers in these countries tend to take up teaching much younger. Nigeria stands alone as having much younger primary school teachers (on the average 28 years old) with an average of about seven years of teaching experience.

TABLE 2.2 Selected characteristics of students by level of science achievement (Population 1)

Percent Mean Science Achievement: Population 1 (Core + 2 Rotated Tests)

Achievement Groups	65 % or more				60 - 65 %			55 - 59 %				50 - 54 %				less than 50 %		
Country	JAP	FIN	KOR	SWE4	CAE	HUN	CAF	ITA	AUS	SWE3	NOR	USA	ENG	POL	SIN	HKO	PHI	ISR	NIG
Selected Mean [%]	66.4	66.0	65.7	62.8	61.7	61.7	61.4	59.2	56.9	55.9	55.6	54.8	53.9	52.5	51.8	50.9	42.3	41.5	35.1
Characteristics																			
Student Characteristics n	7,924	1,600	3,489	1,449	5,104	2,596	2,739	5,156	4,259	1,336	1,305	2,822	3,748	4,390	5,547	5,342	16,851	2,351	944
Mean Age [Years:Months]	10:7	10:10	11.2	10:10	11.1	10.3	11.1	10.5	10.6	9:10	10:11	11:3	10.3	10:11	10:10	10.5	11:1	10.9	12:1
Percent Male Students	51.5	51.7	52.9	50.9	51.0	50.1	51.3	52.7	49.0	50.3	52.6	49.2	51.5	51.0	50.3	50.4	44.1	48.9	54.3
Family Size [b]	2.6	2.5	3.6	2.6	2.8	2.4	2.7	2.6	3.0	2.5	2.2	na	2.6	2.7	3.0	3.2	4.1	3.0	4.3
Father's Education	na	10.0	11.7	9.6	na	10.7	na	11.4	12.3	9.6	na	na	na	11.8	8.1	[8.6]	10.0	4.1	na
Mother's Education	na	10.0	10.8	9.7	na	10.2	na	11.0	11.5	9.6	na	na	na	11.8	6.2	[8.0]	9.8	4.5	na
Number of Books	117	142	98	207	na	183	na	89	235	212	224	178	212	92	97	[100]	62	14	127
Class Size	38	24	55	[23]	na	28	25	na	29	[20]	[20]	24	28	27	37	38	40	32	32
Science as % of Total Instruction Time	10	14	13	6	na	10	5	na	5	na	8	10	3	16	11	5	11	8	na

[a] The United States Mean Science Achievement score is derived from the Core [1M] Test items only.
[b] The questionnaire's highest category was "5 or more children". Hence, for the developing countries this is an underestimate of the family size.
[ ] Data supplied by National Centres from other data sources.
na not available

The Contexts of Science Education

TABLE 2.3 Selected characteristics of teachers teaching science by the level of science achievement (Population 1)

Percent Mean Science Achievement: Population 1 (Core + 2 Rotated Tests)

Achievement Groups	65 % or more					60 - 65 %				55 - 59 %				50 - 54 %					less than 50 %		
Country	JAP	FIN	KOR	SWE4	CAE	HUN	CAF	ITA	AUS	SWE3	NOR	USA	ENG	POL	SIN	HKO	PHI	ISR	NIG
Selected Mean [%]	66.4	66.0	65.7	62.8	61.7	61.7	61.4	59.2	56.9	55.9	55.6	54.8	53.9	52.5	51.8	50.9	42.3	41.5	35.1
Teacher Characteristics n	545	121	560	63	na	97	227	310	819	74	89	123	380	567	232	147	475	96	140
Teacher Salary as % of GNP/capita	193	199	212	122	202	79	202	171	207	122	122	152	184	94	134	169	190	na	na
Percent Male	49.5	50.3	58.3	38.1	na	15.5	22.4	12.9	43.5	2.7	35.0	29.3	39.5	15.0	29.3	37.9	8.5	7.0	74.1
Age in Years	35	42	35	47	na	38	41	45	33	49	41	40	39	38	37	37	41	34	28
% Yrs. Post-Sec. Sci. Educ.	18	14	13	na	na	10	7	4	12	na	8	22	na	46	15	na	3	47	11
Days In-service Training Last Year	1.7	0.6	2.3	1.5	na	3.8	1.3	1.2	0.7	1.3	0.3	1.8	1.0	2.9	1.4	0.9	3.0	3.1	0.6
Years Teach. Experience	14.0	18.4	12.7	na	na	17.1	19.6	20.8	10.7	na	15.1	14.8	13.4	16.3	17.7	14.9	16.5	19.7	6.9
Yrs Post-Sec. Education	3.3	4.1	2.1	2.0	na	2.9	[3.9]	0.9	3.5	2.0	3.3	3.3	3.5	3.8	2.7	1.9	4.5	2.6	1.1
Class size taught	39	19	55	na	na	28	25	18	29	na	na	24	29	28	38	39	38	na	34
Hrs./Week Sci. Teaching	3.5	3.2	3.0	na	na	9.0	1.8	3.4	1.7	na	na	3.2	na	13.3	4.0	2.3	10.1	9.9	2.7
Hrs./Week Teaching	25.8	18.1	21.2	na	na	16.6	21.2	20.1	24.0	na	na	24.5	na	23.0	18.8	18.2	19.4	22.9	14.3
Mean % Time Practical Work	41	17	38	na	na	26	15	13	33	na	na	na	38	21	27	13	40	42	26
% Science in Lab.	45	5	38	na	na	32	4	2	6	na	na	26	11	35	15	3	31	55	45
% Perceiving Lack of Lab for Teaching	36	30	32	na	na	25	69	27	39	na	na	na	na	56	13	47	63	45	80

Note: Canada (Eng.) did not submit any data on teachers; England, Norway, Sweden and the United States dropped many questions from the teacher questionnaire.
na   not applicable

The **class sizes** that teachers reported are similar to those supported by students. The teacher's weekly **teaching load** varies between 14 hours in Nigeria to 26 hours in Japan. Indeed, given that teachers' salaries are a high percentage of the educational budget one wonders why such variation occurs. On a relative basis, in Australia, Japan and United States, **teachers' salaries** match the relatively higher workload. It seems clear that teachers in Poland seem to carry a much greater workload relative to salary [as percent of GNP per capita]. In contrast, it would appear that six countries [Korea, Canada (Fr.), Philippines, Finland, Hong Kong, Italy] draw reasonably high salaries but shoulder only a moderately heavy workload. In 1984, in Poland and Hungary, teachers earned less than the GNP per capita. When comparing the **hours teaching science** per week with the number of **hours taught altogether** then it is clear that in Hungary (9.0 hrs), Israel (9.9 hrs), the Philippines (10.1 hrs), and Poland (13.3 hrs), there are teachers who specialize more in the teaching of science. The percentage of **time spent on practical work** in science and on **working in a laboratory** or classroom with equipment differs. In most countries, there is a relatively high proportion of teachers who perceive the **lack of laboratory facilities** as impeding their teaching.

School Characteristics

School principals responded to a school questionnaire. Table 2.4 presents some general characteristics of the schools in the sample and the principals' views of the locus of decision making about science and about the patterns of science teaching.

The total number of hours of **total instruction per year** varies from 672 hours in Hungary to 1134 hours in Nigeria. The mode is shared by Canada (Eng. and Fr.), and England and Singapore at around 901-1000 hours of total instruction. Given that there is mounting evidence about time invested in learning and achievement, the variation is quite remarkable.

The average **size of school** varies from 195 in Finland to 2068 in Korea. Presumably, the size varies according to national policy and the density of the population. However, not all figures are presented here, and it should be noted that in the two Canadas, Norway, and Poland about 20 percent of the schools were located in communities of less than 1,000 persons. The percentages of schools in the sample coming from communities of more than one million were Australia (19%), Hong Kong (100%), Japan (17%), Korea (20%) and Singapore (100%). In nearly all countries schools were **coeducational**.

The figure of the percentage **male teachers** is for the school as a whole and, hence, differs to some extent from the figure given in Table 2.4 which is based on those teachers teaching science to the grade level tested in this study. It is even clearer from Table 2.3 that the primary grades are taught by women. However, it would appear that it is the men teachers who are called upon more to teach science. The **pupil-teacher ratio** is a proxy measure for the resources of a school. It is not the same as class size but is related to it. Those countries with low pupil-teacher ratios are Hungary (11.8), Italy (13.7), the United States (18.4), Canada (Eng.) (18.5), Finland (18.5) and Poland (19.1). The median number of allocated **hours of instruction for science** varies from one to three at this level. Australia and Hong Kong would appear to offer only a token exposure to science. For the majority of the 16 countries, it varies from two to three hours per week. The percentage **use of the laboratories** or equipped classrooms varies a lot: from one

The Contexts of Science Education

or two percent in Italy and Hong Kong to 77 percent in Hungary, 69 percent in Poland and 58 percent and 52 percent in Japan and Korea respectively. Although not presented in the table, it can be reported that all schools in Finland, Italy, Japan, Nigeria, Norway and Poland have parent-teacher committees but Hong Kong (14 percent of schools) and Singapore (24 percent of schools) have few.

Decisions about science and **patterns of science** teaching are matters of policy and contingent upon the deployment of available teacher resources. In 1984, with the exception of England it was the national or regional authority which determined the range of science subjects to be taught. In 1988, however, England began a national curriculum for its school system. In most countries, the national or regional authorities also decide on the content of what will be taught in science. The exceptions are Australia and England reflecting decentralization in these countries. In some cases, the school board also has a say and in some cases the teachers are also involved. When it comes to the selection of textbooks only in Nigeria, Philippines, Poland and Singapore [to a large extent] are there set texts. But quite unlike these countries, while the multi-media textbooks and teacher guides are produced by the Curriculum Development Institute of Singapore, the principal and teachers are free to choose instructional materials, so as to ensure that the materials chosen are the best suited to the needs of their students. In a large proportion of the cases, these schools prefer the materials designed and published by the Curriculum Development Institute. In Australia, England, Finland, Hong Kong, Hungary, Israel and Italy the decisions are made by the board, principal and/or teachers. In terms of the choice of science equipment, it is typically the teachers and school principals who make the decisions.

Science is predominantly taught by the class teacher. Only the Philippines and Poland tend to have teachers who specialize in the teaching of science. To a small extent this is also the case in Japan (16%). In some schools in all countries, the situation arises where several teachers share the teaching of science to the students in the classes sampled.

Structure of Science Instruction

Before proceeding to consider Population 2 it would be appropriate to present supplementary information on science instruction. This information was specially requested from National Centers. It is set out in Table 2.5.

It will be noted that many countries move from general science to separate subjects at Grade 8 or 9. (However, Finland has separate courses from Grade 4 onwards and Poland teaches biology separately from Grade 4). This is the selection point for admission into the secondary school. Australia (in general, but with notable exceptions), Japan and Papua New Guinea, on the other hand, separate into biology, chemistry and physics only for Grades 11 and above, while Nigeria and Norway introduce science specialization with effect from Grade 10.

Whether taught together or separately many countries cover all fields. Israel has only two science subjects, namely "Life Science" and "Physical Science" at the junior high school level. At Grades 9 and above, the students take three hours per week in each of the three subjects, namely physics, chemistry and biology. In the national case study, Israel reported that usually one of the three hours is

TABLE 2.4   Selected characteristics of schools by level of science achievement (Population 1)

Percent Mean Science Achievement: Population 1 (Core + 2 Rotated Tests)

Achievement Groups	65 % or more						60 - 65 %				55 - 59 %				50 - 54 %				less than 50 %		
Country	JAP	FIN	KOR	SWE4	CAE	HUN	CAF	ITA	AUS	SWE3	NOR	USA	ENG	POL	SIN	HKO	PHI	ISR	NIG
Selected Mean [%]	66.4	66.0	65.7	62.8	61.7	61.7	61.4	59.2	56.9	55.9	55.6	54.8	53.9	52.5	51.8	50.9	42.3	41.5	35.1
Characteristics n	221	106	145	na	215	100	119	119	220	na	89	123	181	199	232	139	475	96	78
School Characteristics																			
Pupil-Teacher Ratio	29.3	18.5	51.8	na	18.5	11.8	21.0	13.7	23.5	na	na	18.4	25.0	19.1	25.6	30.2	32.9	na	28.6
Average Size of School	757	195	2,068	na	274	577	346	518	419	na	na	458	245	614	1201	750	1,204	na	706
Sex of School																			
% Boys only	0	0	0	na	0	0	0	1	3	na	[0]	1	0	0	0	6	0	na	0
% Girls only	0	0	0	na	0	0	0	0	1	na	[0]	1	0	0	6	7	1	na	0
% Co-educational	100	100	100	na	100	100	100	99	96	na	[100]	98	100	100	87	91	99	na	100
Hrs./Yr. Total Instruction	na	760	1,060	na	928	672	881	853	1,043	na	na	1,070	984	na	980	879	na	na	1,134
Hrs./Week Science [median]	3	na	3	na	na	3	1.5	2.5	1	na	na	[3.2]	na	3	3	1	3	2	2.5
% Male Teachers	44	40	52	na	32	22	19	9	31	na	42	15	28	20	29	26	10	12	44
% Use of Laboratories	58	5	52	na	18	77	5	1	6	na	na	18	15	69	19	2	36	na	9
%Patterns of Sci. Teaching[a]																			
by Class Teacher only	84	61	94	na	59	na	82	96	82	na	100	62	68	4	65	2	23	na	86
by Specialist	16	0	2	na	3	na	3	1	3	na	0	6	10	77	3	4	23	na	8
by Several Teachers	0	39	3	na	26	na	5	2	7	na	0	24	17	12	22	86	47	na	6
Science not taught	0	0	1	na	2	na	0	0	0	na	0	0	0	0	0	0	0	na	0
Decisions on:																			
Range of Subjects	N	N	NT	N	NB	N	NB	N	NT	N	N	NBT	PT	N	N	NP	N	NT	N
Science Content	N	N	NT	N	NB	N	NB	N	T	N	NT	NT	PT	N	NT	NT	N	NT	NB
Choice of Text	B	NT	na	T	NT	T	B	T	T	T	BT	T	PT	N	NPT	PT	N	T	N
Choice of Equipment	PT	T	PT	T	PT	T	PT	T	T	T	T	T	T	T	T	PT	NT	T	NB

N National or Regional Authority   P School Principal   [ ] Data provided by National Centres from other data sources.
B School Board   T Teacher(s)

[a] Where percentages do not add to 100, some other pattern of science teaching is also employed.

The Contexts of Science Education

TABLE 2.5 Structure of science instruction in each country

	Grade at which Science is split into separate disciplines (B, C, P)	Science Instruction is at least B, C, P	only one or two subjects	Generally Taught simultaneously	consecutively	Percent of Total Instructional Time devoted to Science		
						Pop 1	Pop 2	Pop 3
Australia	Grade 11	yes		yes		5	15	25
Canada(Eng.)	na	na	na	na		na	na	na
Canada(Fr.)[a]	Grade 9	yes		yes		5	9-18	na
China	Grade 7[b]	yes		yes		na	na	na
England	Year 9	yes		yes		3.3-5	10	50
Finland[c]	Grade 4	yes		yes		14.1	13.3	6.5-19.4
Ghana	na	na		na		na	na	na
Hong Kong[d]	Grade 10	yes		yes		5	10	50-75
Hungary	Grade 6[e]	yes		yes		10	22	10-25
Israel	Grade 9	until Gr.10	Gr.11,12	yes		8	12	17
Italy	na	na		na		na	na	na
Japan	Grade 11	yes		yes		10.3	13.3	na
Korea	Grade 9	yes		yes		12.5	11.7	11.9
Netherlands	Grade 7[f]		yes			na	25	12.5-25
Nigeria	Grade 10	yes		yes		na	100-150[h]	150-200[h]
Norway	Grade 10	yes		yes		8	10	16-50
Papua New Guinea	Grade 11	yes		yes		3	14	40
Philippines	Grade 8	yes			yes	11.1	18.5	na
Poland	Grade 4[g]	yes		yes		16	22.5	30.5
Singapore	Grade 9	yes		yes		11.2	7.5-20	22.2-67
Sweden	Grade 7	yes		yes		6	15	20-23
Thailand	Grade 9	yes		yes		12.3	9.5	23.8
United States	Grade 9	yes			yes	10	20	20
Zimbabwe	Grade 10	yes		yes		na	15	27

na Either not applicable or information not available.
a   In Canada (Fr.), biology and physics are taught in Grade 9; chemistry and physics and, in exceptional circumstances biology in Grade 10; and biology, chemistry, and physics in Grades 11 and 12.
b   In China, chemistry, biology and earth science are available from Grade 7 onwards.
c   In Finland, science at Pop 1 level consists of geography, biology and environmental science. Environmental science is a combination of chemistry and physics. The percentage of instructional time includes geography.
d   In Hong Kong, Grade 10 is the officially prescribed level for science being taught in separate disciplines but unofficially about 50% of schools undertake the separation in Grade 9.
e   In Hungary, science subject specialization replaces the five years of general science (knowledge of surroundings). Physics and biology are taught across Grades 6-8 while chemistry is taught at Grades 7 and 8.
f   In the Netherlands, biology is taught in Grade 7, biology and physics in Grade 8, and either physics and chemistry or chemistry and biology in Grade 9. There are four different types of secondary schools, namely: (a) six years of pre-university, (b) five years higher general education, (c) four year intermediate general education, (d) four year lower vocational education.
g   In Poland, biology is taught separately in Grade 4.
h   Time in minutes. Range of time devoted to science varies according to the type of academic stream.

devoted to laboratory work. In contrast, students in the Netherlands may be studying one or two science subjects at any one time in secondary school depending on their academic stream. All countries indicate that they generally teach all three subjects each year but not necessarily simultaneously, whereas the Philippines and the United States indicate that they teach subjects consecutively. This means that biology may be taught in one year and not in the next and the same for other science subjects, frequently in the order of biology, chemistry and then physics at Grades 10 to 12 respectively. More detail on course offerings and requirements is given in Chapter 7.

Finally, estimates were furnished by National Centers on the percentage of total instructional time devoted to science. In most countries this increases from one population level to the next. Finland is a notable exception to this generalization. However, at Population 3, students who opt to "specialize" in science devote the most amount of time to science (as high as 19.4 percent of total instruction time), which is quite comparable to the majority of countries. At Population 1 level the range is from three percent in Papua New Guinea to 14 percent in Finland. At Population 2 level this ranges from 10 percent in Canada (Fr.), England, Hong Kong, Norway and Thailand to 20 percent or more in four countries, namely: Hungary, the Netherlands, Poland and the United States. At Population 3 level the range of time devoted to science is very wide, ranging from 12 percent in Korea to 50 percent in England and for some students in Norway. In Grade 13 in Hong Kong, some students spent 75 percent of their time studying only science subjects. Countries such as Finland, Hong Kong, Hungary, Netherlands, and Norway show wide variation in the amount of time spent on science, depending on the degree of science specialization in the last few years of schooling.

Population 2

Students

Table 2.6 presents information on selected student variables in Population 2. The balance of **boys** and **girls** enrolled in school is still about fifty-fifty but Ghana, Nigeria, Papua New Guinea and Zimbabwe have 60 to 70 percent boys. It is clear that in these countries more girls than boys have dropped out of school. The Philippines, on the other hand, has 58 percent girls. The picture of the average **number of children** per family is similar to Population 1. Since the **father's** and **mother's education** are highly correlated, the measure presented is only father's education. The caveats mentioned for this variable in the Population 1 results apply here. The developing countries' values are overestimates particularly Nigeria, because of the truncation of the measure. **Class sizes** in Australia, Canada (Fr.), Hungary, Italy, Poland and the United States are in the range of 20 to 30 students per class. Finland has a small average class size (16). Ghana, Nigeria, Singapore and Zimbabwe are in the 30s, while Hong Kong, Japan, the Philippines and Thailand are in the 40s and Korea had, in 1984, an average class size of 65. In all, six countries have larger class sizes in Population 2 than in Population 1. They include Canada (Fr.), Hong Kong, Japan, Korea, Nigeria and Philippines. Korea increased in class size from 55 to 65. It has still the largest class size across all the countries.

The Contexts of Science Education

The total **hours of science instruction** per week was calculated by summing each students' number of hours of science instruction, aggregating this sum for the population and dividing by the number of students. Polish students have a high value of seven hours per week. Ghana, Israel and Korea have over six hours per week. The highest scoring countries (Hungary and Japan) have only four hours per week. Papua New Guinea and Thailand have only three hours per week, and Hong Kong only 2.7 hours per week. The total number of all **hours homework** done varies from a low of 4.4 hours a week for both grades in Sweden to a high of nine or 10 hours per week in Hungary, Italy (Grade 9), Nigeria, Singapore and the United States. The average number of **hours homework in science** ranges from less than one hour a week to about six or seven hours in some countries. In Hungary, the highest scoring country in science achievement, the proportion of science homework to all homework is nearly two-thirds. Finally, the percentage of science students saying that they **do experiments** in class range from 11 in Poland to 65 percent in England, 69 percent in Hong Kong, 62 percent in Thailand, and 91 percent in Zimbabwe. Most countries are in the range of 40-55 percent. Finland (21%), Hungary (22%) and Poland (11%) seem particularly low. It may be speculated that such disparity might be due to translation errors or misinterpretation by the students of the questions asked.

Teachers

Table 2.7 presents the mean national values for selected teacher variables.

At Population 2 level, most countries have 70 to 80 **percent of science teachers** who are men. But China, Finland and Thailand have about the same proportion of **men** and **women**, whereas Hungary, Israel, Italy and Singapore have more women than men science teachers. The Philippines continues to have a very high proportion of female teachers. With the exception of Finland and the Philippines, the proportions of male teachers in Population 2 are substantially more than in Population 1. The largest increase in the proportion of male teachers is in Sweden. The increase was from three percent in Grade 3 and 38 percent in Grade 4 to about 75 percent in Grades 7 and 8. The average **age of teachers** is in the 30 to 45 year range and for the most part the average age is correlated with the number of years of teaching experience. China's and Sweden's teachers are much older. Even so, the Population 2 teachers are marginally younger in age than those in Population 1. It is the developing countries which have teachers with relatively low average **years of teaching experience**. Of the 17 countries with data for both Populations 1 and 2, in nine countries the primary science teachers have more years of teaching experience than the secondary school science teachers. This includes China, Finland, Hong Kong, Hungary, Israel, Italy, the Netherlands, the Philippines and Singapore.

The average **years of post-secondary education** is around four, but the values for Finland and the Netherlands are relatively high (5.3 to 5.7 yrs) and China, Hong Kong and Sweden relatively low (1.5 to 2.8 yrs). In the remaining countries the Population 2 teachers have more post-secondary education than the Population 1 teachers. The percentage of their **post-secondary education spent on science** is relatively high (60 to 70 percent) with the glaring exceptions of the Philippines and Sweden. In the other countries understandably the proportion of

TABLE 2.6 Selected characteristics of students by level of science achievement (Population 2)

Percent Mean Science Achievement: Population 2 (Core + 2 Rotated Tests)

Achievement Groups	65% or more		60 - 65 %						55 - 59 %				
Country	HUN	JAP	NET	CAE	KOR	FIN	SWE8	CHN	ITA9	POL	NOR	AUS	CAF
Selected Mean [%] Characteristics	70.7	66.8	63.7	61.6	61.0	60.3	60.3	60.0	59.8	59.5	59.3	58.8	58.5
Student Characteristics n	2,515	7,610	5,025	5,543	4,522	2,546	1,461	2,806	1,398	4,520	1,420	4,917	2,348
Mean Age [Years:Months]	14.3	14.7	15.6	15.0	15:0	14:10	14:10	15:8	14.8	15:0	15:10	14.5	15:3
Percent Male	49.2	51.6	50.3	48.8	52.9	51.6	51.9	47.5	47.5	48.9	50.8	54.4	47.3
Family Size	2.3	2.6	2.8	3.1	4.0	2.7	2.6	2.0	2.5	2.9	3.0	3.2	2.9
Father's Education	10.8	na	12.2	na	9.5	9.2	9.2	10.0	11.6	11.5	na	11.9	na
No. of Books	na	102	183	na	105	149	210	82	149	107	200	242	na
Class Size	28	42	na	na	65	16	na	na	27	25	23	28	30
Average Hrs. Science Homework/Week	6.3	2.3	6.0	2.5	2.3	2.2	0.9	5.2	4.8	5.2	na	2.0	2.7
Total Hrs. Homework/Wk.	9.0	8.5	8.4	6.4	7.0	5.7	4.4	7.5	10.5	8.6	6.4	6.1	[5.1]
Total Hrs. Science/Week	4.2	4.0	na	na	6.5	4.0	na	5.1	4.0[b]	7.0	na	3.7	na
% Doing Experiments	22	47	31	50	37	21	56	45	29	11	41	54	44
% Instructional Time Science	22	13.3	25	na	11.7	13.3	15	na	na	22.5	10	15	9-18

[a] The United States Mean Science Achievement Score is based only on the Core Test items [2M]. The scores for all the other countries are derived from the 2M scores plus the scores from 2 Rotated Tests.
[b] This figure includes geography together with biology, chemistry and physics.
[ ] Data supplied by National Centres from other data sources.

TABLE 2.6 (ctd.) Selected characteristics of students by level of science achievement (Population 2)

Percent Mean Science Achievement: Population 2 (Core + 2 Rotated Tests)

Achievement Groups	55 - 59 % (continued)							50 - 54 %		less than 50 %			
Country	ISR	THA	SIN	SWE7	ENG	PNG	HKO	USA	ITA8	GHA	ZIM	NIG	PHI
Selected Mean [%] Characteristics	58.5	56.7	56.4	56.0	55.9	55.3	55.0	54.8 [a]	52.4	46.7	42.8	42.2	39.7
Student Characteristics n	2,082	3,780	4,430	1,557	3,118	2,193	4,973	2,519	4,622	2,769	2,698	804	10,871
Mean Age [Years:Months]	14:9	15.4	15:3	13:10	14:2	17:1	14:7	15:3	13:11	16:1	16:1	16:2	16:1
Percent Male	46.4	47.5	50.9	47.9	52.2	63.2	54.8	50.9	52.0	70.4	63.5	62.7	41.9
Family Size	3.8	4.3	3.7	2.6	2.8	4.6	3.7	na	2.7	4.5	4.4	4.5	4.4
Father's Education	na	10.2	7.1	9.1	na	4.3	[7.9]	na	11.2	12.0	10.2	13.6	9.7
No. of Books	na	163	117	200	206	115	[150]	186	114	131	59	200	78
Class Size	na	42	38	na	28	na	40	22	23	39	39	36	48
Average Hrs. Science Homework/Week	3.9	3.6	4.9	1.0	2.1	7.1	[3.1]	6.7	3.7	6.0	5.3	7.4	4.7
Total Hrs. Homework/Wk.	8.4	7.9	9.6	4.4	6.0	7.9	[6.1]	9.6	8.1	8.3	8.1	10.0	7.1
Total Hrs. Science/Week	6.3	3.0	5.0	na	3.0	3.0	2.7	4.4	3.6	6.6	3.6	na	na
% Doing Experiments	50	62	48	51	65	49	69	37	47	50	91	43	54
% Instructional Time Science	12	9.5	7.5-20	15	10	14	10	20	na	na	na	na	18.5

[a] The United States Mean Science Achievement Score is based only on the Core Test items [2M]. The scores for all the other countries are derived from the 2M scores plus the scores from 2 Rotated Tests.
[] Data supplied by National Centres from other data sources.

TABLE 2.7 Selected characteristics of teachers teaching science by level of science achievement (Population 2)

Percent Mean Science Achievement: Population 2 (Core + 2 Rotated Tests)

Achievement Groups	65% or more				60 - 65 %						55 - 59 %				
Country	HUN	JAP	NET	CAE	KOR	FIN	SWE8	CHN	ITA9	POL	NOR	AUS	CAF
Selected Characteristics Mean [%]	70.7	66.8	63.7	61.6	61.0	60.3	60.3	60.0	59.8	59.5	59.3	58.8	58.5
Teacher Characteristics n	354	314	393	na	625	244	94	607	319	798	76	1,630	224
% Male Science Teachers	30	85	89	na	65	49	73	45	39	25	82	74	73
Teacher Salary	79	199	na	na	121	199	149	na	171	94	150	207	202
Age in Years	38	38	37	na	35	39	45	47	42	37	36	34	40
Years Post-Sec. Educ.	4.0	3.8	5.3	na	4.1	5.7	2.0	1.5	4.2	4.1	4.7	4.6	[4.8]
Yrs. Post-Sec. Science Educ.	2.6	2.3	3.0	na	2.6	3.9	0.5	na	3.2	2.4	3.5	2.5	[1.3]
Yrs. Teach. Experience	15.8	15.7	11.6	na	9.8	12.5	na	na	15.9	16.3	11.0	10.4	16.4
Days/Year In-service Training	4.5	3.8	2.0	na	3.2	3.2	2.1	na	2.9	3.6	0.8	1.5	3.2
Hrs. Total Teaching/Week	16.4	18.5	21.4	na	19.8	17.2	23.1	na	17.8	22.4	na	19.3	18.9
Hrs. Science Teach./Week	10.1	16.3	17.2	na	18.0	12.2	15.1	na	12.1	15.8	na	13.8	16.4
% Science of All Teaching	62	88	80	na	91	71	65	na	68	70	na	72	87
Hrs./Week Preparing and Marking	18.4	14.6	15.6	na	15.3	14	na	15	12.4	13.3	na	15.5	11.2
% Science in Laboratory	67	46	67	na	37	72	na	na	23	45	na	77	74
% Time Practical Work	27	32	20	na	52	37	na	17	11	23	na	37	27.0
% Perceive Lack of Equipment	19	46	19	na	34	17	na	46	20	55	na	12	15

Note: Canada (Eng.) did not provide any teacher data. It should also be noted that England, Norway and Sweden did not administer several questions because they were deemed either unacceptable (e.g. infringement of one's privacy) to the teacher, or in the case of China, conceivably unavailable. Teacher Salary is the median yearly teacher salary divided by the GNP per capita. The median teacher salaries were provided by the National Centers.

[ ] Data supplied by National Centers from other data sources.

The Contexts of Science Education 29

TABLE 2.7 (ctd.). Selected characteristics of teachers teaching science by level of science achievement (Population 2)

Percent Mean Science Achievement: Population 2 (Core + 2 Rotated Tests)

Achievement Groups	55 - 59 % *(continued)*							50 - 54 %			less than 50 %			
Country	ISR	THA	SIN	SWE7	ENG	PNG	HKO	USA	ITA8	GHA	ZIM	NIG	PHI
Selected Mean [%]	58.5	56.7	56.4	56.0	55.9	55.3	55.0	54.8	52.4	46.7	42.8	42.2	39.7
Characteristics													
Teacher n	59	96	225	96	1,077	94	133	119	1,425	70	256	261	241
Characteristics													
% Male Science Teachers	30	50	38	76	69	75	66	70	30	93	73	79	10
Teacher Salary	na	192	156	149	193	986	243	149	171	18	na	na	198
Age in Years	35	28	35	45	36	40	31	38	41	30	31	33	33
Years Post-Sec. Educ.	4.8	4.0	3.6	2.0	4.3	3.4	2.8	3.4	4.1	4.2	3.4	4.8	4.8
Yrs. Post-Sec. Science Educ.	3.6	2.5	2.8	0.5	na	1.6	2.3	2.1	2.8	3.3	1.8	3.9	0.8
Yrs. Teach. Experience	10.9	6.5	10.7	na	12.0	14.8	7.7	13.9	14.5	6.3	7.2	9.5	9.8
Days/Year													
In-service Training	4.8	2.2	1.9	1.9	2.2	3.9	2.0	3.0	2.6	1.2	2.0	1.5	4.1
Hrs. Total Teaching/Week	22.7	15.3	17.7	22.6	na	18.2	18.6	23.8	17.0	17.2	19.2	13.7	20.5
Hrs. Science Teach./Week	17.5	13.9	14.0	14.9	na	12.8	12.1	17.4	7.3	14.7	17.1	11.4	16.3
% Science of All Teaching	77	91	79	66	na	71	65	73	43	86	89	83	79
Hrs./Week													
Preparing and Marking	16.5	15.7	21.8	na	na	14.8	15.9	16.7	11.0	17.3	18.3	13.4	13.4
% Science in Laboratory	50	60	52	na	86	78	57	19	13	69	66	55	46
% Time Practical Work	31	49	32	na	42	38	41	na	12	29	na	32	46
% Perceive Lack of													
Equipment	11	28	8	na	na	39	16	na	21	67	35	52	56

Note: Canada (Eng.) did not collect any teacher data. It should also be noted that England, Norway and Sweden did not administer several questions because they were deemed either unacceptable (e.g. infringement of one's privacy) to the teacher or in the case of China, conceivably unavailable. Teacher Salary is the median yearly teacher salary divided by the GNP per capita. The median teacher salaries were provided by the National Centers. The PNG teachers include many Australian teachers recieving Australian salaries.

[] Data supplied by National Centres from other data sources.

post-secondary education in science was much higher for Population 2 teachers when compared to the Population 1 teachers. The average number of **days of in-service training** per year ranges from less than a day a year in Norway to nearly five days a year in Hungary and Israel followed by Japan (3.8 days). Relative to Population 1, in-service training is more frequent for the Population 2 teachers. Most countries have two to three days per year.

The **weekly teaching load** is about 20 hours a week. In a majority of cases, the teachers' workloads are marginally lower than for those in Population 1. Nigeria, as in Population 1, is particularly low. So are Thailand (15.3), Italy (Grade 8) (17.0) and Hungary (16.4). The percentage of the total weekly **load devoted to the teaching of science** is in the 70s and 80s. Italy (Grade 8) is low (43 percent) which, in general, involves the teaching of mathematics and science to the same students in contrast to Korea and Thailand (91 percent) and Zimbabwe, Nigeria, the Netherlands and Japan with a range of 80-89 percent. Teachers were also asked how many hours per week they spent in both **preparing lessons** and **marking homework** or classwork. This ranged from 11 to 22 hours a week but in most countries it was 14 or 15 hours a week.

The percentage of time in science teaching spent on **practical work** ranged from a low of 12 percent in Italy to 52 percent in Korea with Hong Kong, England, the Philippines and Thailand close behind (41, 42, 46 and 49 percent, respectively). The proportion of time spent in laboratories was also very variable ranging from 13 percent in Italy (Grade 8) and, surprisingly, 19% in the United States to 86 percent in England. Other countries in the 70 percent range were Australia, Canada (Fr.), Finland and Papua New Guinea. What is more important than **percent science in laboratories** is whether or not the quality of equipment in the science laboratory permits appropriate concept formation and learning. The percentage of teachers perceiving their science teaching to be hampered by lack of equipment was over 40 percent in China (46%), Ghana (67%), Japan (46%), Nigeria (52%), the Philippines (56%) and Poland (55%). With the exception of Japan the percentage was substantially lower for Population 2 than Population 1 teachers.

Schools

Table 2.8 presents the values for selected variables that characterize the schools from which the students were drawn.

Average **school size** was about 700, but Finland and Hungary were in the 500s and Hong Kong, the Philippines, Singapore and Thailand over 1,000. Most schools in most countries were coeducational. The major exceptions were Korea (only 35% coeducational) and Nigeria. Korea had about equal numbers of boys only, girls only and coeducational schools. Nigeria had 35 percent boys' schools, 13 percent girls' schools and 52 percent coeducational schools. Australia, England, Ghana, Hong Kong, Papua New Guinea and Singapore had from 15 to 26 percent of schools which were not coeducational. This might well reflect an English tradition of educational provision in single-sex church schools.

When considering all teachers in school, most countries had more **male teachers** than female teachers. Hungary and Poland however had 80 percent female teachers. Finland, Italy and Israel also had predominantly more women than men. With the exception of Italy (Grade 9), the Philippines, Singapore and Papua New Guinea, the male **science teachers** represented a higher proportion

The Contexts of Science Education

than **all male teachers** in the school. Nevertheless, Finland, Hungary, Israel, Italy, Papua New Guinea, the Philippines, Poland and Singapore still had more women than men science teachers. Relative to the number of male science teachers at the primary school level, most secondary schools had more male than female science teachers with the exception of Finland, Hungary and Poland. It will be noted that there is a slight discrepancy between Tables 2.7 and 2.8 for the percent male science teachers. The data for Table 2.7 were taken from the responses of the science teachers teaching Population 2 students only, whereas the data for Table 2.8 were reported by school principals for all science teachers in the school.

The average number of **allocated instructional hours** for science as reported by the teachers was different from the number reported by the survey of the students. In general, the figure reported by the school principal was higher than that reported by students. The school principal was presumably reporting time for all science subjects whereas students were studying a particular combination of subjects. The lowest number of hours was in Hong Kong and the highest in Israel, Hungary, Nigeria and Singapore. Most countries had approximately four hours per week.

In terms of the average **number of laboratories** and rooms with science equipment per school, England was the highest (seven) and the lowest was Italy (8) with slightly less than one per school. Most had between two and four laboratories per school. Schools had a high percentage use of the laboratories they possessed; most had a 60 to 80 percent usage. Italy was particularly low (22-38%) and the highest was in England (88%). Most schools had **laboratory assistants**. The range was from no lab assistants per school (Finland and Norway) to 3.0 and 3.3 in the Netherlands and China. The 5.2 **laboratories** or **equipped classrooms** per school in the United States coupled with 79 percent use of labs, but the low number of lab assistants presumably means that the teachers are meant to do what in other countries is done by lab assistants. The United States reported in an internal document that "laboratory assistants or technicians are almost non-existent..... In most schools, the experienced science teacher or Department Head was given one or more periods of time off [release from teaching] to coordinate departmental activities....[such as], order equipment and supplies requisition textbooks and laboratory books and schedule classes and laboratories." The high percentage of the use of laboratories coupled with only 19 percent of the time spent by students in the laboratories would seem to indicate a serious shortage of laboratory space in the United States. Finally, the locus of decision making is with few exceptions similar to the pattern exhibited in Population 1.

Population 3

Population 3 (Science) was defined as those students studying science in courses leading to the study of science at the university level. It will be recalled from Table 1.2 in Chapter 1 that the percentage of an age group of those enrolled, namely: (a) in school; (b) in academic tracks; and (c) in special biology, chemistry or physics courses all vary widely. The average number of subjects they study ranges from three to nine and more and the average age varies from 17 to 19 years.

TABLE 2.8 Selected characteristics of schools by the level of science achievement (Population 2)

Percent Mean Science Achievement: Population 2 (Core + 2 Rotated Tests)

Achievement Groups	65% or more		60 - 65 %						55 - 59 %				
Country	HUN	JAP	NET	CAE	KOR	FIN	SWE8	CHN	ITA9	POL	NOR	AUS	CAF
Selected Mean [%] Characteristics	70.7	66.8	63.7	61.6	61.0	60.3	60.3	60.0	59.8	59.5	59.3	58.8	58.5
School n Characteristics	99	199	224	209	189	90	71	105	72	201	75	233	101
Hrs./Yr. Total Instruction	768	na	1,007	na	1,083	874	na	na	1,094	na	1,110	969	953
% Male Science Teachers	30	86	87	83	67	48	na	50	40	30	81	71	76
% Male Teachers	19	67	79	66	62	36	na	48	48	20	51	55	65
School Size	580	784	612	628	770	506	na	805	774	583	na	760	939
Sex of School:													
% Boys only	0	0	6	0	34	0	[0]	0	3	0	[0]	10	4
% Girls only	0	0	6	3	31	0	[0]	0	2	0	[0]	6	4
% Co-educational	100	100	88	97	35	100	[100]	100	95	100	[100]	84	92
Hrs./Wk. Science [median]	[6.0]	4.0	5.6	na	3.5	3.0	na	na	4.0	na	3.0	3.4	4.9
% Use Laboratories	78	68	66	70	45	82	na	na	38	77	na	82	70
Lab. Equipped Rooms	2.7	2.0	2.7	3.5	1.6	4.4	na	3.9	3.0	2.5	1.5	5.2	5.0
No. of Lab. Asst./School	1.0	na	3.0	1.4	1.1	0.0	na	3.3	1.7	na	0.0	1.5	1.1
% Parent Tch. Committee	na	100	89	na	76	100	na	na	100	100	100	80	[50]
*Decisions on:													
Range of Subjects	N	N	NT	NP	NT	N	N	na	N	N	N	NPT	N
Science Contents	N	N	T	N	NT	N	N	na	T	N	NT	NT	N
Choice of Text	T	B	T	NT	NT	BT	T	na	T	N	BT	T	T
Choice of Equipment	T	PT	T	T	T	T	T	na	T	T	T	T	T

Note: Sweden and Zimbabwe did not collect school data.
[ ] Data supplied by National Centers from other data sources.
* N = National or Regional
B = School Board
P = School Principal or Head of Department
T = Teacher

TABLE 2.8 (ctd.) Selected characteristics of schools by the level of science achievement (Population 2)

Percent Mean Science Achievement: Population 2 (Core + 2 Rotated Tests)

Achievement Groups	55 - 59 % (continued)							50 - 54 %			less than 50 %			
Country	ISR	THA	SIN	SWE7	ENG	PNG	HKO	USA	ITA8	GHA	ZIM	NIG	PHI
Selected Mean [%]	58.5	56.7	56.4	56.0	55.9	55.3	55.0	54.8	52.4	46.7	42.8	42.2	39.7
Characteristics													
School n	58	96	185	222	147	78	110	119	222	53	na	82	272
Characteristics													
Hrs./Yr. Total Instruction	1,420	1,248	1,023	na	1,025	na	1,015	1,141	1,044	1,025	na	1,111	na
% Male Science Teachers	40	53	36	na	68	na	na	64	34	88	na	74	16
% Male Teachers	37	44	41	na	55	66	64	48	34	78	na	70	26
School Size	na	1,113	1,461	na	927	583	1,175	984	470	579	na	813	1,599
Sex of School:													
% Boys only	na	2	11	na	8	8	11	3	1	12	na	35	2
% Girls only	na	4	15	na	7	7	8	2	0	10	na	13	1
% Co-educational	na	94	74	na	85	85	81	95	99	78	na	52	26
Hrs./Wk. Science [median]	7.5	3.0	8.4	na	3.9	3.6	2.4	na	3.8	4.2	na	7.8	5.1
% Use Laboratories	44	57	74	na	88	75	67	79	22	54	na	53	59
Lab. Equipped Rooms	3.5	1.4	4.1	na	7.0	1.9	3.5	5.2	0.9	2.4	na	2.8	3.2
No. of Lab. Asst./School	2.4	0.9	1.2	na	2.1	1.6	2.5	0.7	1.5	2.5	na	2.5	2.0
% Parent Tch. Committee	41	48	10	na	97	73	12	69	100	92	na	84	73
*Decisions on:													
Range of Subjects	N	B	PT	N	PT	N	P	NB	N	N	na	N	N
Science Contents	NT	N	NT	N	T	N	T	BT	T	NT	na	NT	N
Choice of Text	T	NT	T	T	T	NT	T	T	T	T	na	NT	N
Choice of Equipment	T	T	T	T	T	T	T	T	T	T	na	NT	T

Note: Sweden and Zimbabwe did not collect school data.
[ ] Data supplied by National Centers from other data sources.
* N = National or Provincial
B = School Board
P = School Principal
T = Teacher

It is only those who are studying (in a specialized way) biology, chemistry and physics who are discussed in this chapter.[1]

'In a specialized way' is difficult to identify. For example, in Finland 41 percent of an age group is enrolled in academic streams but all must study biology so that all of them are defined as studying biology. Chemistry and physics are optional subjects. However, in Hungary 18 percent of an age group is enrolled in academic streams. All students in the 18 percent study general science but some study extra biology, chemistry, or physics. It was those students studying the extra science who are included in the definitions. A careful perusal of Table 1.2 will help the reader to identify the percentage of an age group involved in the study of biology, chemistry, and physics at the terminal grade. The United States, however, did not test in 1986, as the rest of the world did, all students studying in 12th Grade. Rather they selected students studying the second year of a subject whether it be in 10th, 11th, or 12th Grade. Nevertheless, those studying a second year course should be at the acme of their achievement and this is, then, an élite group for the United States. At Population 3, a total of 17 countries took part in the study. Philippines did not take part because it has an education system that ends at Grade 10. Other countries opted not to test at this level because of lack of financial resources; they preferred to concentrate on Populations 1 and/or 2.

What do the data collected in this study tell us about the Population 3 students, teachers and schools?

Students

Table 2.9 presents the data about the students in each of the subject areas at Population 3 level. Countries are grouped into five groups by levels of scores. The explanation is given in the table. Regrettably, there are many missing data in the tables. This is due to the fact that either the data were not collected or there were high proportions of non response. More **girls** (equal to or more than 70 percent) than **boys** study biology in Canada (Fr.), Japan, Norway, and Poland, but in Ghana, Hong Kong, and Italy many more boys (about 70 percent or more) than girls study the subject. For chemistry, there is a similar pattern without the differences being so pronounced, except in Poland. In Japan, the percentage of boys studying chemistry is 74 whereas for biology it was only 27. Again, with the exceptions of Poland and Thailand physics would appear to be a male subject where 65 percent or more are male students. The trend across most countries appears clear namely, in the study of more specialized science the majority of females opt for biology.

The **number of children per family** within each country is about the same for students across the three subject areas. The range is between two and three children per family with the exceptions of Ghana, Hong Kong, Korea, and Thailand where the range is four to five children per family. The number of years of **father's education** are also quite similar across all three subject areas.

The number of years **further education** which the students report that they expected to have depends in most instances on the number of additional years of education required to pursue a university, college, or polytechnic education. This

---

[1] For a summary of Science Courses by Year Levels, the reader is referred to Volume 1 of this series: *Science Education and Curricula in Twenty-three Countries*. The variations in subject specializations at different grade levels across each country is clearly summarized and reported by the respective National Research Coordinators.

The Contexts of Science Education

turns out to be between three and four years and there are no unexpected findings.

Several countries did not administer the question about the **language used at home** as compared with the language of instruction at school. Singapore has a high percentage of students (ranging from 20 to 40 percent) speaking a different language at home because Singapore has three major ethnic groups: 74 percent Chinese, 12 percent Malay, and 10 percent Indians. What is unique is that English is the medium of instruction across all subjects in school, and students learn two languages (English language and the mother tongue) with effect from Grade 1. Australia has 7.5 percent of students studying biology, who speak a different language at home; 16 percent for chemistry and 17 percent for physics. Canada (Fr.), England, and Thailand are the other countries with four to 14 percent of students speaking a different language at home. It should be of interest for these countries to examine achievement differences for these groups compared with the mother tongue groups in their countries.

Class size is another variable for which there are sparse data. For those countries where the data are reported here, there are large differences between countries and between subjects. Korea (58 students per class) and Thailand (40) both have large class sizes for the teaching of all subjects. Finland, Ghana, Hungary, Singapore, and Poland have class sizes of 25 to 30. England has small class sizes (10 to 12). Hungary, however, has many fewer in chemistry classes (11) than in biology (30) and physics classes (29).

The **hours per week of total science instruction** varies considerably from country to country for biology students. Ghana (the highest at 19 to 20.5 hours), Hong Kong, Hungary, Israel, and Singapore have more than 13 hours per week. Ghana, Hong Kong (16.5 hours), and Singapore (18.6) are particularly high. It would appear that the biology students in these countries are studying two or three science subjects, and hence, are in the highly specialized science-oriented academic streams. For chemistry students, the differences among countries vary from a low of about four hours in Finland and Korea to the high of about 20 hours in Ghana, across the three subjects. The picture for physics students is similar to that for chemistry. It ranges from 5.5 hours in Poland to 19 hours in Ghana.

Finland devotes the least number of **hours** to **biology, physics and chemistry** but Hong Kong, Hungary, and Ghana devote the most hours. In Singapore the total number of hours of science instruction per week decreases from a high of 18.6 hours for the biology stream students to the low of 6.6 hours for physics. This is in part, a result of the way in which the target population was defined (see Appendix B) and, in part, because the biology students, being the science oriented students tend to enrol for all three science subjects. For each of the science subjects, England, Ghana, Hong Kong, Hungary and Singapore provide between five and seven hours per week, while Finland and Korea average about two hours per week only for biology and chemistry.

With the exception of Korea (with an average of 2.2 hours), all countries have an average number of **hours for all homework** per week of between seven and 14 hours. The proportion of all homework per week devoted to doing science homework is an indicator of the amount of time devoted to actual science instruction. In England, Ghana, Hong Kong, Hungary, and Singapore this proportion is high denoting a good deal of effort on the part of these students to science. In contrast Finland, Japan, Korea, the Canadians and Sweden would seem to have a good deal of homework in other subject areas.

TABLE 2.9 *Selected student characteristics by level of science achievement across biology, chemistry and physics samples in Population 3*

Level 1: > 65%
Level 2: 60-64%
Level 3: 55-59%
Level 4: 50-54%
Level 5: < 50%

Categorization of countries across five levels of achievement [1-5], across Biology, Chemistry, and Physics

		Group A				Group B				
		HKO7 (1,1,1)	ENG (1,1,2)	SIN (1,1,3)	HKO6 (1,2,3)	HUN (3,4,3)	JAP (5,3,3)	GHA (3,2,5)	POL (3,5,4)	NOR (3,5,4)
Student Characteristics										
n	Bio	1,559	884	902	2,975	301	1,212	210	764	276
n	Chem	3,599	892	945	5,952	143	1,468	351	763	283
n	Phys	3,690	917	1,071	5,906	398	1,187	291	1716	443
Mean Age [Years/Months]		19:2	18:0	18:1	18:4	18:0	18:2	18:10	18:7	18:11
Science as % of Total Instructional Time		50-75	50	22.2	50-75	10-25	na	na	30.5	16-50
Class Size	Bio	na	10.9	24.4	na	30.2	36.0	15.7	30.5	na
	Chem	na	11.7	24.1	na	10.8	39.7	31.4	30.9	na
	Phys	na	11.7	23.7	na	28.5	39.4	30.6	29.6	[24.0]
Hours per Week	Bio	5.7	5.2	6.3	4.5	5.0	3.1	7.0	4.8	[5.0]
	Chem	5.4	5.2	6.4	4.6	5.2	3.4	6.8	3.4	[5.0]
	Phys	5.4	5.1	6.2	4.7	7.4	3.6	6.9	4.8	[5.0]
Hours Total Science per Week	Bio	16.5	8.3	18.6	13.0	13.5	4.8	20.5	10.2	[5.0]
	Chem	13.2	10.2	12.6	11.3	16.5	6.4	19.5	10.1	[5.0]
	Phys	12.9	8.2	6.6	11.3	11.8	6.8	19.0	5.5	[5.0]
Hrs. Total Homework per Week	Bio	13.4	11.6	14.1	12.3	13.5	9.8	12.2	11.7	8.9
	Chem	13.3	11.4	14.3	12.4	12.9	14.2	12.1	11.7	9.9
	Phys	13.3	11.4	13.9	12.4	12.2	15.1	11.9	11.5	9.6
Hrs. Sci. Homework per Week	Bio	11.2	6.6	9.0	10.2	6.2	2.3	9.1	5.6	na
	Chem	11.1	8.0	9.0	10.1	5.6	4.9	8.8	5.6	na
	Phys	11.1	6.9	5.2	10.1	6.6	5.3	8.8	4.3	na
% Doing Experiments	Bio	100	100	100	81	97	94	100	74	na
	Chem	100	100	99	88	100	98	100	75	na
	Phys	99	99	99	95	87	96	98	75	na
Percent Male	Bio	68	40	51	68	36	27	82	24	26
	Chem	77	67	60	76	36	74	88	24	41
	Phys	78	78	68	77	71	87	89	46	68
Average No. of Children in Family	Bio	3.9	2.7	3.4	3.8	2.1	2.4	4.4	2.4	3.0
	Chem	3.9	2.7	3.5	3.8	2.1	2.5	4.4	2.4	3.0
	Phys	3.9	2.6	3.6	3.8	2.1	2.4	4.5	2.4	3.0
Years Father's Education	Bio	9.9	na	10.4	10.1	12.9	na	12.0	na	na
	Chem	9.8	na	8.5	9.9	11.5	na	11.8	na	na
	Phys	9.7	na	8.9	10.1	13.3	na	11.6	13.3	na
Expected Years of Education	Bio	3.1	2.9	3.9	2.9	3.0	2.3	3.8	3.5	na
	Chem	3.0	3.1	3.7	2.9	1.4	3.0	3.5	3.5	na
	Phys	3.0	2.9	3.7	3.0	3.3	3.0	3.6	3.5	na
% Different Language at Home	Bio	na	6.0	42.5	na	0.7	na	na	0.0	na
	Chem	na	9.1	21.7	na	0.7	na	na	na	na
	Phys	na	7.0	20.5	na	0.5	na	na	0.0	na

Note: Groupings [A-D] by Country Science [Biology, Chemistry, Physics] Achievement
Group A : Countries with Bio., Chem., Physics achievement within levels 1-3. Example: HKO6 (1,2,3) means HKO6's Bio., Chem., Physics achievement is in levels 1, 2, 3 respectively.
Group B : Countries with Bio., Chem., Physics achievement within levels 2-5.
Group C : Countries with Bio., Chem., Physics achievement within levels 4-5.
Group D : Countries with Bio., Chem., Physics achievement within level 5 [less than 50%] only.

TABLE 2.9 (ctd.) Selected student characteristics by level of science achievement across biology, chemistry and physics samples in Population 3

Level 1: > 65% Level 2: 60-64% Level 3: 55-59% Level 4: 50-54% Level 5: < 50%		Categorization of countries across five levels of achievement [1-5], across Biology, Chemistry, and Physics									
		Group C			Group D						
		SWE (4,5,5)	ISR (4,5,4)	FIN (4,5,5)	AUS (5)	CAE (5)	ITA (5)	KOR (5)	THA (5)	USA (5)	CAF (5)
Student Characteristics											
n	Bio	619	879	1652	1631	3254	147	3319	1171	659	249
n	Chem	1172	243	971	1177	2923	217	2979	1168	537	1187
n	Phys	1156	472	810	1073	2766	1766	981	1161	485	944
Mean Age [Years/Months]		19:0	17:7	18:7	17:3	18:3	19:0	17:11	18:3	17:7	17:2
Science as % of Total Instructional Time		20-23	17	6.5-19.4	25	na	na	11.9	23.8	20	na
Class Size	Bio	[21.9]	na	25.9	19.2	na	na	58.2	41.0	17.3	na
	Chem	24.0	na	22.2	18.0	na	na	58.8	[24.0]	na	na
	Phys	24.0	na	21.9	17.4	na	na	58.3	40.9	17.3	na
Hours per Week	Bio	[2.0]	na	[2.0]	4.3	na	[3.2]	2.1	3.1	na	na
	Chem	1.3	na	[1.0]	4.1	na	[3.1]	1.9	[3.1]	na	na
	Phys	[3.8]	na	[2.0]	4.2	na	3.2	3.9	4.0	na	na
Hours Total Science per Week	Bio	[5.5]	15.4	[4.0]	5.3	na	[6.4]	6.8	10.1	na	na
	Chem	na	na	[4.4]	8.2	na	[5.4]	4.4	[5.5]	na	na
	Phys	[5.5]	na	[4.7]	8.0	na	6.4	7.1	10.1	na	na
Hrs. Total Homework per Week	Bio	9.4	8.7	9.1	13.4	7.7	8.6	2.3	7.7	9.4	7.5
	Chem	8.0	9.2	9.4	14.5	8.2	9.4	2.2	7.4	10.9	6.6
	Phys	[8.0]	9.6	8.9	13.9	8.0	10.3	2.4	7.6	12.7	6.0
Hrs. Sci. Homework per Week	Bio	3.5	na	2.8	4.1	3.2	5.3	0.8	4.6	4.9	3.2
	Chem	3.6	na	3.3	6.3	3.5	6.4	0.8	4.5	na	2.7
	Phys	3.5	na	3.4	5.9	3.6	5.2	1.1	4.6	5.5	2.6
% Doing Experiments	Bio	95	na	13	98	98	65	72	98	95	99
	Chem	98	na	23	99	99	88	72	na	na	98
	Phys	92	na	24	99	99	76	70	99	93	95
Percent Male	Bio	56	40	47	33	37	82	52	46	41	26
	Chem	70	51	48	67	53	69	52	45	59	44
	Phys	70	69	63	69	70	81	75	48	78	51
Average No. of Children in Family	Bio	2.5	3.2	2.7	3.2	3.3	2.7	4.1	4.4	na	3.0
	Chem	2.5	3.0	2.8	3.3	3.2	2.6	4.1	4.4	na	3.0
	Phys	2.6	2.8	2.7	3.2	3.2	2.6	4.0	4.4	na	2.9
Years Father's Education	Bio	10.7	13.2	9.4	12.3	na	na	10.4	10.3	14.9	na
	Chem	10.2	14.4	9.3	13.0	na	na	10.3	10.3	14.9	na
	Phys	10.3	13.7	9.7	12.8	na	na	10.5	10.4	15.7	na
Expected Years of Education	Bio	na	3.2	3.2	2.7	na	2.1	2.9	3.1	3.2	4.3
	Chem	na	3.4	3.3	3.1	na	2.2	3.0	3.1	3.6	3.4
	Phys	na	3.3	3.4	3.1	na	2.7	3.1	3.1	3.7	3.5
% Different Language at Home	Bio	4.5	na	0.5	7.5	na	0.0	na	8.2	na	14.4
	Chem	2.8	na	0.6	15.9	na	na	na	9.3	na	11.4
	Phys	2.6	na	0.6	17.0	na	na	na	9.2	na	9.0

Note: Groupings [A-D] by Country Science [Biology, Chemistry, Physics] Achievement
Group A : Countries with Bio., Chem., Physics achievement within levels 1-3. Example: HKO6 (1,2,3) means HKO6's Bio., Chem., Physics achievement is in levels 1, 2, 3 respectively.
Group B : Countries with Bio., Chem., Physics achievement within levels 2-5.
Group C : Countries with Bio., Chem., Physics achievement within levels 4-5.
Group D : Countries with Bio., Chem., Physics achievement within level 5 [less than 50%] only.

Finally, the percentage **doing experiments** is on the whole high with the glaring exception of Finland which is consistently low, ranging from 13 percent to 24 percent across the three science subjects.

Teachers

Table 2.10 presents data on selected characteristics of the teachers who teach biology, chemistry, and physics at the Population 3 level.

Where countries did not collect teacher data, this is indicated as not available (na). This includes Canada (Eng.), Israel, Norway, and Sweden. Papua New Guinea is included even though it did not administer the achievement tests to the biology, chemistry, and physics specialist groups of science students.

In general, chemistry and physics are taught by **male teachers**. More women teach biology, than chemistry and physics in most countries. In Poland, 88 percent of the biology teachers are women. In contrast, in England, Ghana, Hong Kong, Japan and Korea, most biology teachers are men. Hungary, Italy, Singapore, and Thailand also have a majority of female biology teachers. The phenomenon is consistent with the overall trend in schools.

In general, the average **age** of all teachers is between 32 and 43 years. However, in Ghana and Hong Kong the average age of teachers is 28 and 32 respectively. Relative to the primary school teachers, these teachers are much younger.

On average, all the teachers had at least three to six **years of post-secondary education**. This is consistent with the fact that they are graduate teachers who possess a three to four year degree followed by another year or so of teacher certification. Almost all spent at least two thirds of their post secondary education studying science. In some countries, the teachers spent as much as 80 percent of their time or more at the post secondary level studying science. These countries were Finland, Ghana, Hong Kong, Hungary, Italy, and Singapore. It is an indication of the level of specialization that is expected for science teaching at the Population 3 level.

There are large differences between countries in terms of **years teaching experience**. For biology teaching the Canadian (Fr.), Finnish, Hungarian, Japanese, Polish, and United States teachers all had more than 16 years teaching experience but Hong Kong and Thai teachers had fewer than nine years experience. This is also reflected among the physics and chemistry teachers, with a significant variation for Canada (Fr.) (12.6 years) and England (14.2 years).

In about nine out of 16 countries, teachers receive 3.5 to 4.5 days of **inservice training** per year but in two of the subjects Ghana, and Singapore have less than two days. In an extreme case, physics teachers in Ghana had an average of seven days a year of in-service but only 0.9 days for biology. Hong Kong has more in-service training for biology than for chemistry and physics. The Singapore teachers received on the average 0.6 of a day in-service only.

The total **teaching load per week** also varied from a low of 13 hours per week (in physics in Thailand) to over 21 hours in Canada (Fr.), Poland, and the United States. Most teachers had an average of 17 or 18 hours per week of teaching. Teachers reported spending 15 to 20 hours per week in **preparing lessons and marking students' assignments**, both during and after school.

However, teachers in Canada (Fr.) and Italy would seem to spend much less time on these two activities than their counterparts in other countries. Both spent consistently around nine to 11 hours across all three science subjects. It would appear that teachers in Hungary, Japan and Singapore devote consistently more time (20 to 25 hours per week) to preparing lessons and marking students' work than teachers in other countries.

The time teachers allocate to **practical work** in their teaching ranged widely across countries. The values recorded for Italy (7 percent) and Japan (15 percent) are surprisingly low in the teaching of physics practical work. Australia, England, Hong Kong, Singapore, and Thailand have high figures compared with other countries in biology.

On the other hand, the percentage of all instruction taking place in **laboratories** is relatively high with the exception of Italy (23 to 34 percent).

The percentage of teachers in each country which thought that the **lack of laboratory facilities** hampered their teaching of science was, in general, low for biology but with unexpectedly high figures for Poland in chemistry (54 %), Japan in biology and physics (38%) and Korea in biology (36%). Papua New Guinea teachers were unanimous about the lack of science equipment for physics teaching and a majority of Polish teachers were concerned about the lack of chemistry equipment and perceived this to be a severe source of limitation to their science teaching. About a quarter to a third of the science teachers in Korea and Japan perceived the lack of equipment to be a severe limitation to their science teaching. This is surprisingly high when it has been the general impression that their schools are well equipped with laboratory facilities. Singaporean teachers reported that they perceived there were no severe limitations in laboratory equipment for the teaching of biology and chemistry.

In general, the picture is one of hard working teachers with varying amounts of preservice teacher training. All have some in-service training. They have a work load of about 17 hours per week but spend another 16 hours preparing and marking student work. About 30 to 40 percent of time is spent on practical work and about 40 to 80 percent teaching takes place in laboratories. A high number of Korean and Japanese teachers perceived themselves to be severely hampered in their science teaching due to a lack of equipment, but in the majority of countries only about 5 to 15 percent of teachers perceived this as a shortcoming.

Schools

Table 2.11 presents the data on selected characteristics of schools.

On the number of **hours total instruction**, five of the 12 countries had on the average more than 1000 hours per year. This ranged from about 1086 hours (Italy) to 1346 hours for Israel, with United States, Norway, Thailand, and Singapore falling in between. For the remainder, they cluster around 900 to 1000 hours per year.

The average **size of schools** is about 950 students per school. The range is from about 388 in Hungary and Norway to nearly 2000 in Thailand. Poland (488) and Finland (535) also have relatively small school sizes. Korean classes were reported to be the largest (around 58 students per class), and the average school size is only 772. These schools consisted of Grades 10, 11, and 12 only.

TABLE 2.10 Selected teacher characteristics by level of science achievement across biology, chemistry and physics samples in Population 3

Level 1: > 65%
Level 2: 60-64%
Level 3: 55-59%
Level 4: 50-54%
Level 5: < 50%

Categorization of countries across five levels of achievement [1-5], across Biology, Chemistry, and Physics

		Group A				Group B			
		HKO7 (1,1,1)	ENG (1,1,2)	SIN (1,1,3)	HKO6 (1,2,3)	HUN (3,4,3)	JAP (5,3,3)	GHA (3,2,5)	POL (3,5,4)	NOR (3,5,4)
Teacher Characteristics										
n	Bio	162	248	8	200	84	40	5	145	na
n	Chem	203	259	10	251	na	46	3	60	na
n	Phys	199	283	16	245	89	37	2	164	na
Percent Male	Bio	55	66	0.0	59	38	83	80	12	na
	Chem	82	82	0.0	83	na	98	50	23	na
	Phys	94	86	39	94	62	95	100	38	na
Age in Years	Bio	32	37	37	32	42	40	32	41	na
	Chem	31	38	34	31	na	43	29	40	na
	Phys	31	39	32	31	40	42	28	40	na
Years Teaching	Bio	8.5	13.0	12.7	8.7	17.9	16.1	na	18.9	na
Experience	Chem	8.5	14.2	8.7	8.3	na	19.0	4.7	18.0	na
	Phys	7.9	14.2	7.1	7.7	15.4	18.1	3.0	18.3	na
Years Post-Sec.	Bio	4.0	4.7	4.0	3.9	5.4	4.4	3.8	5.5	na
Education	Chem	4.0	4.8	3.0	4.0	na	4.3	4.3	5.6	na
	Phys	3.9	4.5	4.0	3.8	5.4	4.2	5.0	5.2	na
% Post Secondary	Bio	86	na	87	86	82	61	87	78	na
Science Educ.	Chem	87	na	87	87	na	62	86	75	na
	Phys	87	na	87	87	79	65	87	72	na
Days of In-service	Bio	4.0	na	2.4	3.8	3.9	3.6	0.9	3.6	na
Training	Chem	3.3	2.2	0.9	3.4	na	3.9	1.3	3.4	na
	Phys	3.2	2.8	0.1	3.1	4.3	3.6	7.0	3.4	na
Hours/Week	Bio	21.3	na	25	20.6	21.1	19.2	21.7	15.9	na
Preparation	Chem	19.4	na	20.4	19.0	na	21.4	16.7	13.7	na
and Marking	Phys	19.3	na	22.8	19.3	22.1	21.7	13.0	16.1	na
Hours Total	Bio	17.7	na	14.9	17.3	14.7	15.7	15.0	21.9	na
Teach. per Week	Chem	17.4	na	15.6	17.5	na	15.7	15.7	20.7	na
	Phys	17.1	na	16.0	17.2	17.0	15.9	15.5	22.8	na
% of Time	Bio	36	35	46	36	24	19	31	26	na
Practical Work	Chem	37	35	38	36	na	22	50	29	na
	Phys	31	32	39	30	21	15	50	22	na
% Science in Lab	Bio	57	82	64	57	57	43	75	49	na
	Chem	54	82	56	53	na	42	70	53	na
	Phys	52	84	55	52	67	47	90	45	na
% Lack of	Bio	12	na	0	16	24	38	0	41	na
Equipment	Chem	12	na	0	13	na	20	0	54	na
	Phys	18	na	13	20	16	38	0	29	na

Note: Papua New Guinea did not administer the biology, chemistry, and physics tests at Population 3.

Groupings [A-D] by Country Science [Biology, Chemistry, Physics] Achievement:
Group A : Countries with Bio., Chem., Physics achievement within levels 1-3. Example: HKO6 (1,2,3) means HKO6's Bio., Chem., Physics achievement is in levels 1, 2, 3 respectively.
Group B : Countries with Bio., Chem., Physics achievement within levels 2-5.
Group C : Countries with Bio., Chem., Physics achievement within levels 4-5.
Group D : Countries with Bio., Chem., Physics achievement within level 5 [less than 50%] only.

TABLE 2.10 (ctd.) Selected teacher characteristics by level of science achievement across biology, chemistry and physics samples in Population 3

Categorization of countries across five levels of achievement [1-5], across Biology, Chemistry, and Physics

Level 1: > 65%
Level 2: 60-64%
Level 3: 55-59%
Level 4: 50-54%
Level 5: < 50%

		Group C			Group D						
		SWE (4,5,5)	ISR (4,5,4)	FIN (4,5,5)	AUS (5)	CAE (5)	ITA (5)	KOR (5)	THA (5)	USA (5)	CAF (5)	PNG -
Teacher Characteristics												
n	Bio	na	na	31	276	na	26	217	151	43	17	6
n	Chem	na	na	21	221	na	188	208	140	37	24	7
n	Phys	na	na	32	203	na	321	166	140	34	20	7
Percent Male	Bio	na	na	55	55	na	39	84	35	51	77	66
	Chem	na	na	52	80	na	41	90	46	65	80	71
	Phys	na	na	75	87	na	48	96	84	94	95	86
Age in Years	Bio	na	na	45	36	na	39	38	32	43	39	34
	Chem	na	na	36	36	na	43	38	31	43	37	35
	Phys	na	na	41	35	na	41	38	32	43	43	35
Years Teaching	Bio	na	na	18.6	11.2	na	11.9	11.9	8.6	16.7	16.1	9.5
Experience	Chem	na	na	11.8	13.1	na	15.5	11.7	7.9	17.2	12.5	10.7
	Phys	na	na	14.1	12.5	na	15.7	12.3	7.8	16.7	18.6	12.7
Years Post-Sec.	Bio	na	na	5.9	4.8	na	4.0	4.4	4.4	5.9	5.8	5.0
Education	Chem	na	na	5.8	5.0	na	4.5	4.5	4.4	5.8	5.6	4.9
	Phys	na	na	5.9	4.9	na	4.2	4.4	4.5	6.4	5.5	4.6
% Post Secondary	Bio	na	na	87	67	na	87	65	66	71	62	75
Science Educ.	Chem	na	na	65	71	na	84	72	67	74	75	71
	Phys	na	na	68	69	na	68	72	64	71	63	82
Days of In-service	Bio	na	na	3.6	2.0	na	2.9	2.1	4.4	3.1	3.7	2.8
Training	Chem	na	na	3.4	2.2	na	3.6	2.2	3.5	3.7	3.4	3.4
	Phys	na	na	3.3	1.7	na	3.1	2.1	4.1	3.1	3.5	4.3
Hours/Week	Bio	na	na	12.2	17.3	na	10.0	15.0	15.5	16.8	8.8	15.2
Preparation	Chem	na	na	17.8	17.9	na	11.6	15.4	15.3	18.9	10.9	13.5
and Marking	Phys	na	na	18.7	17.2	na	11.9	16.7	15.6	17.0	9.4	13.0
Hours Total	Bio	na	na	15.9	18.6	na	18.0	18.2	13.7	21.7	23.5	18.6
Teach. per Week	Chem	na	na	16.9	17.3	na	18.7	18.0	13.7	21.0	21.2	19.3
	Phys	na	na	16.1	16.7	na	17.7	17.2	13.1	21.7	21.9	16.6
% of Time	Bio	na	na	17	37	na	22	na	44	na	28	17
Practical Work	Chem	na	na	50	32	na	18	na	50	na	26	32
	Phys	na	na	31	30	na	7	na	48	na	25	39
% Science in Lab	Bio	na	na	na	83	na	32	na	61	88	73	72
	Chem	na	na	na	83	na	34	na	73	83	85	80
	Phys	na	na	na	82	na	23	na	65	87	77	90
% Lack of	Bio	na	na	10	11	na	12	36	23	na	12	na
Equipment	Chem	na	na	5	12	na	22	27	21	na	15	0
	Phys	na	na	16	12	na	14	35	35	na	15	100

Note: Papua New Guinea did not administer the biology, chemistry, and physics tests at Population 3.

Groupings [A-D] by Country Science [Biology, Chemistry, Physics] Achievement:
Group A :   Countries with Bio., Chem., Physics achievement within levels 1-3. Example: HKO6 (1,2,3) means HKO6's Bio., Chem., Physics achievement is in levels 1, 2, 3 respectively.
Group B :   Countries with Bio., Chem., Physics achievement within levels 2-5.
Group C :   Countries with Bio., Chem., Physics achievement within levels 4-5.
Group D :   Countries with Bio., Chem., Physics achievement within level 5 [less than 50%] only.

TABLE 2.11 Selected school characteristics by level of science achievement across biology, chemistry and physics samples in Population 3

Level 1: > 65%
Level 2: 60-64%
Level 3: 55-59%
Level 4: 50-54%
Level 5: < 50%

Categorization of countries across five levels of achievement [1-5], across Biology, Chemistry, and Physics

	Group A				Group B				
	HKO7 (1,1,1)	ENG (1,1,2)	SIN (1,1,3)	HKO6 (1,2,3)	HUN (3,4,3)	JAP (5,3,3)	GHA (3,2,5)	POL (3,5,4)	NOR (3,5,4)
School Characteristics									
n	115	127	16	158	76	193	na	150	153
Size of School	1,237	908	1,508	1,179	388	1,074	na	488	377
% Male Science Teachers	71	70	43	71	47	93	na	na	[84]
% Male Teachers	52	59	42	53	40	80	na	33	[68]
Sex of School:									
% Boys only	15	18	0	12	0	5	na	0	0
% Girls only	8	9	0	7	0	17	na	0	0
% Coeducational	77	73	100	81	100	78	na	100	100
Hours/Year Total Instruction	983	942	1,272	980	na	na	na	na	1,110
Hours/Week Allocated to... Bio	6.1	4.9	6.3	5.4	[5.0]	2.8	na	[4.0]	[5.0]
Chem	6.0	4.9	6.3	5.4	[5.2]	3.2	na	[3.0]	[5.0]
Phys	6.0	4.9	6.3	5.4	[2.4]	3.1	na	[5.0]	[5.0]
Hrs./Week Total Science Instruction Time	17.6	15.9	12.1	16.1	[8.0]	10.0	na	12.0	na
Average No. of Labs per School	4.3	8.8	7.3	4.3	3.5	4.4	na	5.1	na
% Use of Labs	76	86	71	75	80	53	na	86	na
Average No. of Lab. Assistants	2.8	2.8	5.0	2.7	1.3	1.4	na	na	na

Note: Papua New Guinea did not administer the biology, chemistry, and physics tests at Population 3.

Groupings [A-D] by Country Science [Biology, Chemistry, Physics] Achievement:
Group A : Countries with Bio., Chem., Physics achievement within levels 1-3. Example: HKO6 (1,2,3) means HKO6's Bio., Chem., Physics achievement is in levels 1, 2, 3 respectively.
Group B : Countries with Bio., Chem., Physics achievement within levels 2-5.
Group C : Countries with Bio., Chem., Physics achievement within levels 4-5.
Group D : Countries with Bio., Chem., Physics achievement within level 5 [less than 50%] only.

TABLE 2.11 (ctd.) Selected school characteristics by level of science achievement across biology, chemistry and physics samples in Population 3

Level 1: > 65%
Level 2: 60-64%
Level 3: 55-59%
Level 4: 50-54%
Level 5: < 50%

Categorization of countries across five levels of achievement [1-5], across Biology, Chemistry, and Physics

	Group C			Group D						
	SWE (4,5,5)	ISR (4,5,4)	FIN (4,5,5)	AUS (5)	CAE (5)	ITA (5)	KOR (5)	THA (5)	USA (5)	CAF (5)	PNG -

School Characteristics											
n	na	48	86	165	370	317	207	160	118	107	4
Size of School	na	na	535	816	798	845	772	1984	1205	1036	329
% Male Science Teachers	na	42	63	66	85	39	85	47	63	75	80
% Male Teachers	na	37	37	52	57	47	80	41	49	62	70
Sex of School:											
% Boys only	na	na	0	14	1	3	48	4	1	0	0
% Girls only	na	na	0	17	2	0	38	2	4	7	0
% Coeducational	na	na	100	69	97	97	14	94	95	93	100
Hours/Year Total Instruction	na	1346	na	983	992	1086	na	1203	1160	950	na
Hours/Week Allocated to... Bio	na	3.0	[2.0]	[4.0]	na	0.5	1.5	3.0	na	1.7	na
Chem	na	4.6	[1.0]	[4.0]	na	2.0	1.8	3.0	na	3.2	na
Phys	na	4.7	[2.0]	[4.0]	na	2.1	2.2	4.0	na	3.5	na
Hrs./Week Total Science Instruction Time	na	14.7	[5.0]	14.1	na	5.5	7.0	12.1	na	7.2	8.0
Average No. of Labs per School	na	3.7	3.1	5.9	4.7	2.9	2.4	5.9	5.9	5.3	3.8
% Use of Labs	na	44	72	83	75	39	34	74	82	73	72
Average No. of Lab. Assistants	na	2.6	na	1.6	1.4	1.8	1.1	1.9	1.2	1.1	1.0

Note: Papua New Guinea did not administer the biology, chemistry, and physics tests at Population 3.

Groupings [A-D] by Country Science [Biology, Chemistry, Physics] Achievement:
Group A : Countries with Bio., Chem., Physics achievement within levels 1-3. Example: HKO6 (1,2,3) means HKO6's Bio., Chem., Physics achievement is in levels 1, 2, 3 respectively.
Group B : Countries with Bio., Chem., Physics achievement within levels 2-5.
Group C : Countries with Bio., Chem., Physics achievement within levels 4-5.
Group D : Countries with Bio., Chem., Physics achievement within level 5 [less than 50%] only.

Most systems have **coeducational schools**. Only Australia, England, Hong Kong, and Japan have 20 to 30 percent of schools which are single-sex schools. Korea has overwhelmingly single-sex schools and only 14 percent are coeducational.

Finland, Hungary, Israel, Poland, Singapore, and Thailand have more female than male teachers in their schools at this level. However, about 12 out of 17 systems of education are dominated by male science teachers. The exception to this is Italy where the proportion of female to male teachers is higher for science teachers than for the schools as a whole.

The **number of hours** per week allocated for each of the **science subjects within** each country is relatively uniform. However, when countries are compared Hong Kong and Singapore provided the most number of instructional hours per week with an average of six hours per subject. It is significant to note that those countries with a higher amount of instructional time per subject (Singapore, Hong Kong, and England) are those who have performed significantly better than all the other countries at Population 3 level.

The average **number of laboratories** per school is nearly five per school with England having 8.8 and Korea having only 2.4. The percentage use made of the laboratories is 70 to 80 percent with the notable exceptions of Italy and Korea where only 39 and 34 percent made use of the laboratories. This is confirmed by the Korean teachers who reported only about 25 percent use of the laboratories for science teaching. Japan also is worthy of mention with only 53 percent of science teaching time being held in laboratories, while Israel is not far behind at 44 percent. All schools have one or two laboratory technicians per school but England, Hong Kong, and Israel have nearly three per school while Singapore has about five per school, on the average. In Singapore, the reliance on the laboratory assistant is due partly to the specialized nature of science subject teaching in laboratories at the junior colleges or pre-university classes.

Conclusions

The conclusions to this chapter emphasize the major differences amongst the systems.

Percentage of an Age Group in School

At Population 1 level nearly all the systems studied had all children enrolled in school. The exception was the Philippines where five percent of boys had already dropped out of school. Population 2 was the last point in school in most systems where all or nearly all children were still in school. However, China, Ghana, Papua New Guinea, the Philippines, Thailand, and Zimbabwe had experienced a drop out of children of 40 to 89 percent. By Population 3 level only, Japan, Korea, and the United States had 80 to 90 percent of an age group still enrolled in school. Canada had about 70 percent, Australia had 39 percent, Finland 63 percent, Hungary 40 percent, Israel 65 percent, and Norway 40 percent. The others had anywhere from one percent (Ghana) to 34 percent (Italy). However, the proportion of an age group actually studying a science subject in the academic courses in the last year of secondary school was from three to 41 percent in biology, one to 37 percent in chemistry and one to 35 percent in physics.

Family Background

In the developing countries, students were typically from large families and from homes with fewer resources than students in the developed countries. In nearly all countries there was a selection factor between Populations 2 and 3 with students from "better" homes going on to the final year of secondary school science courses.

Sex of Science Teacher

At Population 1, most teachers were female but Korea and Nigeria had more male teachers. At Population 2, however, most science teachers were male. But, China, Finland and Thailand had about the same proportion of men and women science teachers whereas in Hungary, Israel, Italy, Singapore and particularly the Philippines there were more women than men science teachers. This situation continued at Population 3 for Italy, Poland and Singapore where there were more female than male science teachers. In general, more women taught biology than men but there were some exceptions to this general statement. However, the chemistry and physics teachers were predominantly male.

Time Devoted to Learning Science in School

At Population 1, science instructional time as a percentage of total instructional time varied from three to 16 percent. At Population 2, it varied from ten to 25 percent and at Population 3 level it varied from seven to 75 percent. At Population 3 this indicates the amount of specialization in science since, in some countries, a science subject can be one of three or four subjects and, in other cases, one of nine or more subjects. The 75 percent case was where students spent 75 percent of their time on science subjects. At Population 2 the actual average number of hours per week varied from three hours in Thailand to 10 hours in Italy (Grade 9) and at Population 3 it varied from about four to 18 hours for each of the branches of science.

The percentage of time in science involving practical work varied from 13 to 40 percent at Population 1 level, and the percentage of time involving experiments varied from 30 to 90 percent at Population 2 and seven to 50 percent at Population 3 level. Japan reported that while a high proportion used the laboratory, only a small percentage of students 'do experiments'. This is due to the fact that teachers conduct brief laboratory demonstrations without the students having hands-on experimentation. However, Hong Kong obviously has a lack of laboratories and in some countries the laboratories are not in full use (see Table 2.4).

Homework

At Population 2, students did about eight hours per week (range six to 10 hours) of homework for all subjects together. The proportion of that time spent on science homework ranged from about one quarter to three quarters. In most systems, however, students spend between two and four hours per week on science homework.

At Population 3, total homework lay between seven and 15 hours with the exception of Korea where it was only two hours. The average was about 10 hours per week. The average time given to science homework was about 11 hours per week.

Teacher Training

At Population 1, most teachers had had about three years of preservice training. However, in Nigeria this was only one year. The Nigerian teachers were also the youngest among all the countries and were mostly non-graduates. The percentage of that time spent on science education ranged from three to 47 percent. Most countries lay between 10 and 20 percent.

At Population 2, most educational systems had an average of four to five years of pre-service education. The exception were China and Sweden with 1.5 years and two years respectively. The proportion of that time given to science education was between one quarter and a half.

However, at Population 3, teachers had an average of four to six years of pre-service education and the percentage of that time given to science education was about 70 percent. Thus, the science teachers of the final grade of secondary school were, in general, well trained - at least, in terms of time - for their jobs.

Systems of education varied a great deal in the average number of days per year inservice training. At Population 1, this varied from an average of 0.3 days (Norway) to 3.8 days (Hungary). At Population 2, it varied from 0.8 days to 4.5 days for the same two countries. Again, at Population 3, Hungary had the highest number of days (four to five depending on the branch of science). The lowest recorded figure was for Physics teachers in Singapore (0.1 days per year).

Number of Hours Teaching per Week

At Population 1, the number of total hours of teaching per week ranged from 14 hours in Nigeria to nearly 26 hours in Japan. Most countries were in the range of 18 to 23 hours per week. Most teachers taught science for two to four hours per week but in Hungary, Poland, the Philippines, and Israel this was nine to 13 hours per week indicating the use of special science teachers in the primary grades.

At Population 2, the total teaching load ranged from 14 hours (again in Nigeria) to nearly 24 hours in the United States. In most countries the load was about 18 or 19 hours per week. The number of hours those teachers spent teaching science ranged from 12 to 18 hours per week. In most countries it was between 11 and 14 hours per week. The proportion of the total teaching load devoted to science teaching was a half to two thirds. The glaring exception was

The Contexts of Science Education

Grade 8 in Italy where it was only two fifths. In this case, however, the student is generally taught mathematics and science by the same teacher.

At Population 3 level, the total teaching load was 16 to 21 hours and nearly all of this time was spent teaching science subjects. This confirms that the science teachers were specialized according to their different subject disciplines.

Hours Spent by Teachers on Preparing Lessons and Marking Homework

At Population 2, the average number of hours devoted by teachers to preparing lessons and marking homework was 11 to 22 hours. In most countries, the range was 13 to 15 hours. This was the equivalent of about 70 to 100 percent of the time spent on teaching.

At Population 3, the number of hours spent on preparation and marking ranged from nine to 25 hours. The proportion of preparation and marking to teaching load ranged from under half in Canada (Fr.) to nearly one and a half in Singapore. It was in the higher achieving countries that more time was spent on preparing and marking than on actual teaching.

Curriculum

Most countries had a national curriculum or nationally prescribed syllabus. However, at Population 1 this was not true for Australia and England, where teachers had the major say on the context they taught in science. In most countries, however, the teachers had a large say in the textbooks they used and the equipment the school acquired.

At Population 2, the situation was a little different. The teachers in Australia, England, Italy and the Netherlands had a large say in the content they taught. In systems like Hong Kong and Singapore, the centralized decisions about the science content took into account the views of teachers who were represented on the Ministry's curriculum development committees. The extent of the teachers' autonomy was different from that of England where the teachers had a direct say in what was being taught at the school level. For the choice of textbooks and equipment the situation was much the same as for Population 1.

At Population 3, there was either a national curriculum or an external examination system that influenced what the teachers taught.

No information is given in this chapter on courses offered and courses which were compulsory. These varied between and within countries. The patterns of offerings and requirements could, in fact, turn out to be important in influencing student achievement in science as will be considered in later chapters.

School Size

School size (total enrolment in schools) varied considerably. At Population 1 most countries were in the range 400 to 700 students. But Korea, Singapore and the Philippines had average school sizes of 1,200 to 2,000 students. To a great measure, this is a reflection of the population density within the city and town limits wherein most of the school-age children are in schools. Finland, on the other hand, had an average school enrolment of 194 students.

At Population 2, most countries were in the range 500 to 800 students but with Hong Kong and the Philippines having 1,200 and 1,600 students respectively.

At Population 3, the range was 400 to 500 in Hungary, Poland, Finland, and Papua New Guinea but 2,000 in Thailand and in the range 1,000 to 1,500 in Hong Kong, Singapore, Japan and the United States

Single-Sex Versus Coeducational Schools

At Population 1, nearly all countries had coeducational schools. The exceptions were Hong Kong and Singapore where some 13 percent of an age group attended single-sex schools.

At Population 2, the number of systems having students in single-sex schools increased. Many had five to 15 percent single-sex schools but in Korea 34 percent were boys-only schools and 31 percent girls-only schools. In Singapore, 11 percent were boys-only and 15 percent girls-only schools.

At Population 3, Singapore, Hungary, Poland, Norway, Sweden, Finland and Papua New Guinea had only coeducational schools. Those with more than 20 percent of the single-sex schools were Hong Kong (Form 7), England, Japan, Australia, and Korea (only 14 percent in coeducational schools).

Teacher Perception of the Lack of Science Equipment

In the developing countries, fairly high proportions of teachers felt that the lack of equipment hampered them in their science teaching. Perhaps the same extent of complaint can be levelled at the lack of or non-existence of laboratory assistants to help the science teachers.

The above conclusions are based on selected variables. It is for the reader to peruse the tables in the chapter if he or she is interested in other variables. It can be seen that there are, indeed, differences in the way in which educational systems are organized and, in particular, how they organize their science education. The reader should bear in mind that such differences do not of themselves imply differences in school effectiveness. The purpose of drawing attention to such variations in the contexts of science education is to note that such diversity reflects differences in traditions, priorities and national educational policies and goals for schooling and science education.

Chapter 3 looks in detail at the cognitive achievement of the students.

Reference

1. Thorndike, R. L. 1973 *Reading Comprehension Education in Fifteen Countries.* Almquist and Wiksell, Stockholm.

3

Science Achievement

In the preceding chapter, the background variables were presented in terms of how nations performed on the cognitive tests. The aim of this chapter is to present briefly the cognitive science scores of countries and comment on them. Before examining the various scores it is first desirable however to examine the construction of the achievement measures.

Measures of Science Achievement

The tests were constructed on the basis of the common intended curriculum in all of the participating countries. The intended curriculum is that content which is included in national or state syllabi, the major science textbooks used by students, and - where applicable - the national examinations. A first analysis had been conducted in the late 1960s for the First IEA Science Study. This was repeated for the second science study. Sufficient items had to be the same for the first and second science studies to allow comparisons between the two times of testing. The final composition of the items was:

	Population 1	Population 2	Population 3
Biology	22	23	39
Chemistry	5	15	39
Earth Science	8	9	-
Physics	21	23	38
Total	56	70	
Test 3E			30
Test 3X			30
Test 3N			30

The items for Population 1 were split into a common (core) test of 24 items and four rotated tests of eight items each. Each student took the core test and two rotated tests thus making a maximum score of 40. For Population 2, there was a common test of 30 items and four rotated tests of 10 items each. Hence, a maximum score of 50 was possible. For Population 3 there was a common test containing biology (9 items), chemistry (9 items), and physics (8 items) as well as a special biology test (30 items), a chemistry test (30 items), and a physics test (30 items). The appropriate items were combined to make a total biology or chemistry or physics score. A more detailed description of the tests together with information on the validities of the tests are presented in Volume I of this series. The reliability coefficients (KR-20) of each of these tests is presented in Appendix G in this volume.

It should be pointed out that although the testing was meant to take place in the eighth month of the school year there was, in fact, great variation in the times when systems tested. This ranged from the third to the eighth month.

The Results

Item Results

The item statistics for each item are obtainable on diskette upon request from the IEA-Headquarters (see Preface).

Three examples only of items are given below in order that the reader can see examples of items for different populations and the type of information available for each and every item on the diskettes. The first item presented is from Population 1, the second from Population 2 and the third from Population 3 (Physics). These items are in the same format as on the diskettes available from the IEA-Headquarters.

In the examples given below (and on the diskettes containing all items) the identification data are as follows. The item number indicates the population, the test, and the position of the item within the test. Thus, in the first item below, P1M06 denotes Population 1 [P1], the core test [M], and the sixth item [06]. The content classification identifies Biology [B], Chemistry [C], Earth Science [E], or Physics [P] as the primary area, followed by a two-digit integer between 01 and 57. denoting the subarea. These subareas are identified in detail in Volume I. The content code for the first item below, E03, identifies Earth Science [E], and the subarea of Meteorology [03] within it. The primary data files for the study are organized by population and data source. Response information for items can be found by identifying the particular file within which they are recorded. All of the item response data for students is to be found on the student raw data file for the relevant population. The Test Location (TLOC) denotes the sequential position of an item on that raw data file. The formats for these files can be found in the International Codebook for long form files. In this instance, the code 026 indicates that the responses to the first item listed below may be found in position 26 of the relevant file.

The column identifications within the table for each item are labelled: A, B, C, D, E, Omit., Gender, Diff. The first five of these reference the five multiple choice alternatives for each test item. Within the body of the table, the percentages of response to each alternative are given by country. The asterisk denotes the correct answer. Omit. references the omitted responses. Gender designates the difference between the percentage of correct responses given by males and females. Diff. indicates whether males or females had the larger percentage.

The correct answer to the first item presented is D. The range of correct answers is from 14.6 percent of the students in the Philippines to 70.6 percent in Japan. In general, the 10-year-olds found this to be a difficult question. In no country was there sign of random guessing as could be seen by the frequent equal distribution of students selecting the wrong answers. In eight of the countries slightly more females than males selected the correct answer.

The second item presented is for Population 2 and is a chemistry item. The correct answer is A and it, too, is a relatively difficult item for the 14-year-olds. Only in Hungary do more girls than boys select the correct answer. The most popular distractors are B and C. Most students know that the particle involved is an electron but are unsure of the position. The high percentage of students omitting the item in Italy and England is consistent with the instruction not to guess at random.

Population 1
Item No.   Content Classification   Test Location
P1M06   E03   026

The following question refers to the table below which shows some temperature readings made at different times on three days.

	6 a.m.	9 a.m.	12 noon	3 p.m.	6 p.m.
Monday	15 °C	17 °C	20 °C	21 °C	19 °C
Tuesday	15 °C	15 °C	15 °C	10 °C	9 °C
Wednesday	8 °C	10 °C	14 °C	14 °C	13 °C

On one day a cool wind began to blow. When do you think this happened?

- A   Monday morning
- B   Monday afternoon
- C   Tuesday morning
- D   Tuesday afternoon
- E   Wednesday afternoon

Country	A	B	C	D*	E	Omit.	Gender Difference
Australia	17.8	10.3	11.0	37.1	21.7	2.2	1.6 F
Canada (Eng.)	16.4	12.0	8.7	32.0	28.0	2.7	1.2 F
Canada (Fr.)	14.5	8.9	8.1	46.3	20.1	2.4	0.1 M
England	17.4	9.9	11.2	34.6	22.2	4.7	3.2 M
Finland	13.4	13.7	10.0	41.8	20.5	0.8	0.6 M
Hong Kong	27.4	7.4	19.3	29.6	15.3	1.1	1.1 F
Hungary	14.2	8.6	6.6	55.9	12.5	2.1	4.5 F
Israel	8.6	9.0	7.8	55.4	12.8	6.4	1.7 M
Italy	20.0	12.1	11.3	41.8	10.2	4.7	0.1 M
Japan	5.5	6.2	6.4	70.6	11.0	0.3	1.9 F
Korea	13.1	10.3	6.1	56.5	13.9	0.1	6.3 M
Nigeria	31.7	15.9	15.4	17.6	16.6	3.0	5.7 F
Norway	15.7	23.0	14.5	27.7	15.4	3.8	6.5 M
Philippines	42.1	16.6	10.9	14.6	14.8	1.0	1.5 M
Poland	18.9	12.6	12.5	32.0	18.6	5.5	1.3 F
Singapore	36.0	8.3	14.1	22.4	18.9	0.3	0.2 F
Sweden (Grade 3)	8.9	17.6	7.0	40.4	17.0	9.1	2.1 M
Sweden (Grade 4)	5.2	14.6	2.9	54.1	18.1	5.1	2.7 M
U.S.A. (Phase 2)	24.0	14.3	10.3	25.9	24.8	0.8	2.7 M

* = correct answer

Population 2		
Item No.	Contents Classification	Test Location
P2M20	C33	040

Which of the following particles are gained, lost or shared during chemical changes?

- A electrons furthest from the nucleus of the atom
- B electrons closest to the nucleus of the atom
- C electrons from the nucleus of the atom
- D protons from the nucleus of the atom
- E neutrons from the nucleus of the atom

Country	A*	B	C	D	E	Omit.	Gender Difference
Australia	32.1	17.2	19.8	11.4	8.7	10.8	10.1 M
Canada (Eng.)	40.3	14.1	17.3	11.3	7.2	9.4	10.2 M
Canada (Fr.)	27.1	20.8	19.4	8.0	9.4	15.0	6.4 M
China	56.2	20.4	10.9	6.5	4.0	2.0	11.7 M
England	26.6	19.3	16.8	11.8	7.8	17.8	9.6 M
Finland	37.0	15.0	26.9	14.2	4.8	2.0	1.9 M
Ghana	47.8	13.3	17.8	11.4	8.5	1.3	7.7 M
Hong Kong	16.1	19.1	31.3	21.3	10.5	1.6	6.3 M
Hungary	76.0	8.8	6.2	4.6	2.5	1.8	1.9 F
Israel	48.2	13.8	16.3	6.8	3.5	10.4	10.7 M
Italy (Grade 6)	32.3	14.4	14.3	6.9	8.1	23.9	5.4 M
Italy (Grade 7)	35.9	13.6	7.8	2.7	4.5	35.3	4.5 M
Japan	20.3	25.7	33.9	9.4	9.1	1.5	7.6 M
Korea	16.2	16.4	40.1	11.5	15.7	0.2	7.7 M
Netherlands	47.0	15.0	19.5	8.1	5.9	4.2	0.5 M
Nigeria	36.8	13.7	30.1	9.6	6.9	2.9	2.2 M
Norway	29.3	14.3	24.6	13.3	8.9	9.6	5.9 M
Papua New Guinea	20.7	23.9	27.3	15.8	10.4	2.0	0.9 M
Philippines	35.6	23.9	17.3	12.7	9.9	0.6	1.2 M
Poland	50.9	19.2	15.5	3.9	3.5	6.9	3.4 M
Singapore	32.7	16.1	23.6	13.4	11.1	3.2	9.7 M
Sweden (Grade 7)	36.1	10.3	26.4	12.1	6.0	7.7	5.6 M
Sweden (Grade 8)	43.5	12.1	25.1	9.3	4.6	5.2	7.5 M
Thailand	15.4	41.3	25.2	9.8	7.8	0.5	3.8 M
U.S.A. (Phase 2)	39.4	15.0	20.1	15.4	8.0	1.7	10.1 M
Zimbabwe	17.0	26.5	18.6	12.4	13.6	12.0	4.2 M

* = correct answer

Science Achievement 53

Population 3		
Item No.	Contents Classification	Test Location
P3M23	P40	043

A steel ball rolls down an inclined plane. Which of the graphs below best represents the relationship between the distance travelled (s) and the time (t)? (Assume retarding forces are negligible.)

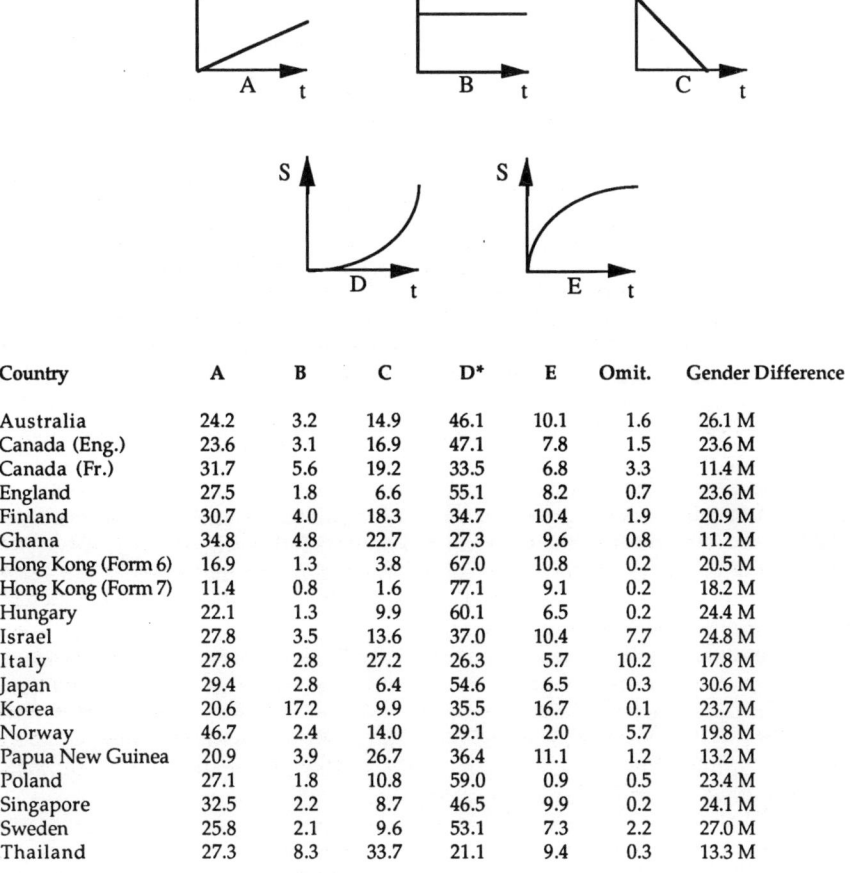

Country	A	B	C	D*	E	Omit.	Gender Difference
Australia	24.2	3.2	14.9	46.1	10.1	1.6	26.1 M
Canada (Eng.)	23.6	3.1	16.9	47.1	7.8	1.5	23.6 M
Canada (Fr.)	31.7	5.6	19.2	33.5	6.8	3.3	11.4 M
England	27.5	1.8	6.6	55.1	8.2	0.7	23.6 M
Finland	30.7	4.0	18.3	34.7	10.4	1.9	20.9 M
Ghana	34.8	4.8	22.7	27.3	9.6	0.8	11.2 M
Hong Kong (Form 6)	16.9	1.3	3.8	67.0	10.8	0.2	20.5 M
Hong Kong (Form 7)	11.4	0.8	1.6	77.1	9.1	0.2	18.2 M
Hungary	22.1	1.3	9.9	60.1	6.5	0.2	24.4 M
Israel	27.8	3.5	13.6	37.0	10.4	7.7	24.8 M
Italy	27.8	2.8	27.2	26.3	5.7	10.2	17.8 M
Japan	29.4	2.8	6.4	54.6	6.5	0.3	30.6 M
Korea	20.6	17.2	9.9	35.5	16.7	0.1	23.7 M
Norway	46.7	2.4	14.0	29.1	2.0	5.7	19.8 M
Papua New Guinea	20.9	3.9	26.7	36.4	11.1	1.2	13.2 M
Poland	27.1	1.8	10.8	59.0	0.9	0.5	23.4 M
Singapore	32.5	2.2	8.7	46.5	9.9	0.2	24.1 M
Sweden	25.8	2.1	9.6	53.1	7.3	2.2	27.0 M
Thailand	27.3	8.3	33.7	21.1	9.4	0.3	13.3 M

* = correct answer

The third item presented is a physics item for Population 3 students. The correct answer is D. In general, it is an item of average difficulty although fewer than 30 percent of students in Ghana, Italy, Norway and Thailand selected the correct answer. The highest percentage of students selecting the correct answer is from Hong Kong.

On this physics item, it can be seen that in all countries boys scored much better than girls, with a difference in favor of boys ranging from 11.2 percent in Ghana to 30.6 percent in Japan.

Achievement Scores

This section reports the total and subscores for each population in turn. Most emphasis is given to Population 2 because it is this group of youngsters where nearly 100 percent of an age group is in school at a point before they leave school to enter the labour market.

As has already been explained, there were 56 items at Population 1 level and 70 items at Population 2 level. At Population 3 level there were 39 items for biology, 39 items for chemistry, and 38 items for physics. These are not many items to cover most aspects of science meant to be learnt in each of the countries at these age levels. Unfortunately, to have used more items would have required more testing time in schools and this was not possible in many of the countries in the study.

Some words of warning should be expressed about the use of the total and subscores presented in this chapter. Different systems place different emphases on different topics in science. With only a 40, 50 or 70 item test the items may not truly reflect the totality of science objectives in each of the educational systems. It can happen that if more items, or different items, had been used a reversal of the rank order of countries could occur where the means presented are very near each other. It should also be pointed out that no correction for guessing of student scores was undertaken.

Throughout the forthcoming presentation of results, it is seen that most are presented twice: once <u>without</u> the United States on a larger number of items (core test plus two rotated tests) and a second time <u>with</u> the United States on a smaller number of items (core test only). This is because, in 1986, the United States did not administer the rotated tests at Population 1 and 2 levels and did not administer the whole of the core test at Population 3 level.

The rank order correlations between country scores on the core test and core plus two rotated tests at Population 1 is .98 and at Population 2 level it is .94. However, the scores based on the core plus two rotated tests have greater reliability and validity than the scores based only on the core test.

National scores were computed in the following way. The correct items for a student were summed and divided by the number of items presented to the student; they were then converted to percentage scores. These scores were then aggregated to the school level, weighted, and further aggregated to the national level. Percentage scores have been used throughout this volume for two reasons. The first is that there is one common metric which makes it easier for the reader. The second is that in some countries for some tests in some populations, the odd item was dropped because of translation errors in the item in a particular country. Standard errors of sampling (ses) are also recorded in percentage form.

Four other tests were also administered. One was a practical test the results of which are being reported in a later publication (Doran and Tamir, 1991). One was a 30 item test (Test 3E) of earth science for Population 3. Only a few countries administered this test. A further test was a general science test of 30 items (Test 3X) and this was taken by only three countries. Finally, there was a 30 item test

(Test 3N) developed for those students not specializing in science at Population 3 level. The Population 3N scores have been used in Figure 3.1 and are presented in Appendix D. The test reliabilities (KR-20) and standard errors of the means of the core tests for Populations 1 and 2 and for the core test scores, biology scores, chemistry scores, and physics scores at Population 3 level are presented in Appendix E.

Population 1

Table 3.1 presents the total scores for each country. The total scores are based on the weighted aggregate of the scores for each of the individuals on both the core and the core plus rotated tests.

TABLE 3.1 *Means and standard deviations for core test and core plus 2 rotated tests (in percent), ROH and percentage of schools scoring lower than lowest scoring school in highest scoring country for Population 1*

Country	N	Core (N=24 Items)					Core +2 rotated tests (N=40 items)		
		Mean	sd	se	ROH	% schools below 52.5	Mean	sd	se
Australia	4,259	53.5	18.8	.73	.15	37	56.9	17.2	.68
Canada (Eng.)	5,104	57.2	17.8	.54	.12	25	61.7	15.8	.48
Canada (Fr.)	2,739	60.4	17.5	.67	.11	12	61.4	15.4	.60
England	3,748	48.8	18.6	.70	.17	61	53.8	17.0	.64
Finland	1,600	63.8	16.8	.62	.07	7	66.0	14.9	.55
Hong Kong [b]	5,342	46.6	17.6	.81	.27	77	50.9	15.8	.72
Hungary	2,590	60.2	18.8	.96	.22	21	61.7	16.3	.85
Israel [a]	2,351	49.6	17.8	1.00	.22	64	41.5	14.7	.76
Italy	5,156	55.8	19.4	1.06	.19	38	59.2	17.6	1.00
Japan	7,924	64.3	16.5	.28	.04	0	66.4	14.7	.24
Korea	3,489	64.0	17.7	.66	.16	7	65.7	16.0	.59
Nigeria	944	32.9	17.3	1.65	.69	88	35.1	15.6	1.53
Norway	1,305	52.9	17.0	1.25	.15	58	55.5	16.3	1.14
Philippines	16,851	39.6	18.7	.66	.56	83	42.3	16.7	.59
Poland	4,390	49.7	18.7	.66	.22	66	52.5	17.2	.62
Singapore [b]	5,547	46.8	17.0	.74	na	75	51.8	16.0	.73
Sweden (Grade 3)	1,336	53.4	17.7	.89	.13	43	55.9	15.9	.78
Sweden (Grade 4)	1,449	61.1	16.6	.65	.03	3	62.8	14.6	.60
U.S.A.	2,822	54.8	19.0	.75	.14	38	-	-	-

[a] The data from Israel are unweighted
[b] The % schools in Hong Kong and Singapore are % of classes.

Since the core plus rotated tests have more items than the core test alone, it is on core plus rotated tests that comparisons are made. It will be noted that the percentage correct is slightly higher on the core plus two rotated score than on the core score only. This is because two of the rotated tests were easier than the core test and the other rotated tests. When examining the scores reference should also be made to the mean ages of each of the samples in Table 1.1 in Chapter 1.

Japan, Finland and Korea perform well at this level closely followed by the two Canadas, Hungary and Sweden. Israel, Nigeria and the Philippines have low scores compared with other countries. The remaining countries are bunched in the 50 to 60 score point range on the core plus two rotated tests. Australia, England, Italy, and Poland have slightly larger standard deviations than other countries. The se column presents the standard error of the means due to sampling variation. It will be noted that the se of the core plus two rotated is slightly less than for the core alone. This is due to additional items rendering more stability in the estimates of school means. In general the se is between 0.5 and 0.8, i.e. between one half and eight-tenths of one percentage point. Only in Italy, Nigeria, and Norway do the standard errors exceed ±five percent of the student standard deviation.

The ROH (Kish, 1965) is an index associated with the ratio of the between group (school) variance to the overall total (student) variance. However, in this study many countries selected one class in each school. Since it is unlikely that students are assigned to classes at random in any school, the variance component between classes within schools is positive. This implies that the within class variation is systematically smaller than the within school variation. Hence, in most countries the estimated between school variation is too large and the coefficient ROH is overestimated. However, in some countries the sampled students came from all classes at a grade level within the school and in these countries the ratio estimates the fraction which the between schools variation is of the overall variation. These latter countries were Australia, Canada (Fr.), England, Ghana, and Poland. The value of .15 for Australia indicates that approximately 15 percent of the variation of scores is between schools and 85 percent within schools. It is to be noted that Sweden (Grade 4), Japan, and Finland have a small ROH, indicating that nearly all the variation is within classes. This, in turn, indicates that there is very little difference between classes or schools so that it makes very little difference for achievement in science which class or school a student attends. However, in Nigeria and the Philippines the science achievement of a student has a great deal to do with which class (or school) the student attends.

As already mentioned, the index measures the between school or class variation relative to total variation. This implies that the ROH is large when either the between school/class variation is large or the within school/class variation is low.

The column headed **"% schools below 52.5"** indicates the percentage of schools or classes in the samples in each country scoring lower than the mean achievement in the lowest class in the highest scoring country - Japan The surprises are England, Hong Kong, Israel, Norway, Poland and Singapore, which have quite high proportions of low scoring schools and classes. It must be recalled that Hong Kong and Singapore sampled classes and not schools. In this case, it is the percentage of classes scoring below the lowest scoring class in Japan.

The tests for Population 1 consisted of a core test of 24 items which was taken by every student, plus four rotated tests of eight items each. It was planned that each student would take the core plus two of the rotated tests. In practice, most countries followed the plan but some chose other patterns of

Science Achievement

administration. The full set of 56 items could be classified in various ways. In this volume the items were classified to render the following subscores: biology (22 items), chemistry (5 items), earth science (8 items), physics (21 items); and separately for information (20 items), comprehension (18 items), and application (18 items). Table 3.2 presents the full achievement test results based on all 56 items for all of these countries - (except the United States which did not administer the rotated tests. Subscores were not calculated for the United States because they were judged to be of insufficient accuracy due to the small number of items.

In biology, Finland and Japan do well whereas Norway and Singapore are not performing much better than Israel, Nigeria, and Philippines which score lowest.

In chemistry, it is Hungary, Finland, and Italy which do well while Japan performs relatively poorly. Extremely poor scores are achieved by students in Australia, England, Hong Kong, and again Israel, Nigeria, and Philippines.

In earth science, (8 items only) Hungary, Japan, and Sweden (Grade 4) all do well whereas in physics it is Japan and Korea which score highest. The performance on the earth science subtest of England, Israel, Italy, Norway, Singapore, and Sweden (Grade 3) is poor but Hong Kong, Nigeria, Philippines, and Poland score lowest.

It is to be noted that many countries have profiles where they have lower scores in comprehension than in information and, again, lower in application than in comprehension. However, this does not hold true for Canada (Fr.), Hungary, Israel, Japan, Korea, Poland and Sweden (Grade 4). In Table 3.2, Israel is in the lowest achievement category with respect to each of the science subject areas but for process areas, Israel is in the lowest category for information and application, but is in the highest for comprehension. This implies exceptionally high performance on comprehension items, but exceptionally low performance in all other areas. In fact, the low performance must be low enough that the subject area averages are lowered to the lowest level even after accounting for the high comprehension performance.

One question of interest is whether or not these relative achievement scores, in any way, reflect achievement scores higher up the school system.

Population 2

The reader should recall that there is considerable variability in the proportion of students in school among the countries at the level of Population 2. One would speculate that this proportion means something different depending on whether it is used to assess the state of the system or to interpret the meaning of the achievement test scores. Low values indicate that the educational system is not as well developed as in countries with high values. This implies that in the competition for resources to extend education, the instruction of those in school may suffer. Also, in these countries, the home backgrounds of students may not prepare them for school learning as well as those in developed countries. On the other hand, countries with smaller proportions of children in school are likely to

TABLE 3.2 Country scores and sub-scores* for Population 1

Score range	Core+2rotated	Biology (22 items)	Chemistry (5 items)	Earth Science (8 items)	Physics (21 items)	Information (20 items)	Comprehension (18 items)	Application (18 items)
more than 65%	FIN JAP KOR	FIN JAP KOR	FIN HUN ITA	HUN JAP SW4	JAP KOR	CAE FIN	FIN JAP KOR	JAP
60-65%	CAE CAF HUN SW4	CAE CAF HUN ITA SW4	CAF KOR NOR SW4	CAE CAF FIN ITA KOR SW3	CAE CAF FIN	AUS CAF HUN ITA JAP KOR NOR SW4	CAE CAF HUN SWE4	CAF FIN HUN KOR SW4
55-59%	AUS ITA NOR SW3	AUS ENG HKO POL SW3	CAE POL SW3	AUS ENG NOR	AUS HUN SW4	ENG SW3	AUS ITA SIN SW3	CAE ITA
50-54%	ENG HKO POL SIN	NOR SIN	JAP	HKO POL SIN	ENG ITA NOR SIN SW3	HKO POL SIN	ENG HKO NOR POL	AUS NOR POL SW3
less than 50%	ISR NIG PHI	ISR NIG PHI	AUS ENG HKO ISR NIG PHI SIN	ISR NIG PHI	HKO ISR NIG PHI POL	ISR NIG PHI	ISR NIG PHI	ENG HKO ISR NIG PHI SIN
	Median 56.4%	Median 58.7%	Median 55.6%	Median 59.7%	Median 53.4%	Median 60.1%	Median 58.4%	Median 52.9%

AUS - Australia; CAE - Canada (English speaking); CAF - Canada (French speaking); CHN - China; ENG - England; FIN - Finland; GHA - Ghana; HKO - Hong Kong; HUN - Hungary; ISR - Israel; ITA - Italy; JAP - Japan; KOR - Korea; NET - Netherlands; NIG - Nigeria; NOR - Norway; PNG - Papua New Guinea; PHI - Philippines; POL - Poland; SIN - Singapore; SW3/SW4 - Sweden; THA - Thailand; ZIM - Zimbabwe.

* The actual percentage subscores from which this table is derived are given in Appendices F.1 and F.2.

Science Achievement

represent the urban, better prepared subpopulations in attendance. This would imply that even if the instructional quality remained the same but attendance increased - scores would go down. In other words, selection effects associated with school attendance increase scores.

The countries with the lowest percentage of the age group in school at the Population 2 level - with grade levels in brackets - are:

Country	Grade level	Percentage of an age group in school
Papua New Guinea	10	11
Zimbabwe	9	30
Thailand	9	32
China	9	37
Ghana	9	43
Nigeria	10	unknown
Philippines	9	60
Italy	9	72
Canada (Fr.)	9	82

All of the other countries exceed 90 percent in school attendance for Population 2. Most of the low percentage enrollment countries have relatively low achievement in comparison to high percentage enrollment countries, indicating less developed educational systems. There are two exceptions, however. These are China and Thailand. In these countries, the selection factors predominate and the selected students perform as well or better than students in some developed countries.

Another important difference among Population 2 countries and sub-populations is the grade level tested. Depending on how the instruction in the country is organized and parcelled out, grade levels can provide an index of the cumulative amounts of science instruction that students have received up to the time of testing. For most of Population 2, the focal grade level was 9. However, several countries tested all or a subgroup of students at a different grade. These were:

Country	Grade level
Sweden (Grade 7)	7
Australia (Grades 8, 9, 10)	8, 9, 10
Finland	8
Hong Kong	8
Hungary	8
Italy (Grade 8)	8
Poland	8
Sweden (Grade 8)	8
Italy (Grade 9)	9
Nigeria	10
Papua New Guinea	10

Here, the lower grades of testing may put some of the developed countries at a disadvantage compared with other developed countries.

Table 3.3 presents the national core and core plus rotated test scores for Population 2. It will be noted that more countries participated in Population 2 than in Population 1. The ROH values and the percentages of schools (or classes) falling below the class with the lowest mean in the highest scoring country

(Hungary) are also presented for the core test only, again because the United States did not administer the rotated tests.

In terms of a total score Hungary and Japan have high scores and the developing countries of Ghana, Nigeria, the Philippines, and Zimbabwe have low scores. The rank order correlation on the core test between those countries common to Population 1 and Population 2 is .84 indicating that the achievement at the national level at Population 2 is very much associated with the science already learned in the primary school grades. Countries having relatively little science at Population 1 level may wish to consider this finding.

TABLE 3.3 Means and standard deviations for core test and core plus 2 rotated (in percent), ROH and percentage of schools scoring lower than lowest scoring school in highest scoring country for Population 2

Country	N	Core (N=30 Items)					Core +2 rotated tests (N=50 items)		
		Mean	sd	se	ROH	% schools below 49.4	Mean	sd	se
Australia	4,917	59.5	16.4	.63	.17	8	58.8	16.0	.62
Canada (Eng.)	5,543	61.9	15.6	.56	.14	6	61.6	14.9	.58
Canada (Fr.)	2,348	60.2	13.9	.67	.16	6	58.5	13.5	.69
China	2,806	58.7	15.9	.90	.22	8	60.0	15.1	.85
England	3,118	55.8	16.4	.73	.19	19	55.9	15.7	.72
Finland	2,546	61.7	13.8	.43	.05	2	60.3	13.9	.44
Ghana	2,769	45.5	16.2	1.20	.48	64	46.7	15.6	1.19
Hong Kong	4,973	54.6	14.9	.83	.29	26*	55.0	14.4	.88
Hungary	2,515	72.2	15.7	.85	.26	0	70.7	15.1	.81
Israel	2,082	61.9	16.5	1.14	.31	12	58.5	17.0	1.25
Italy (Grade 8)	4,622	52.4	16.4	.71	.27	36	52.4	15.3	.66
Italy (Grade 9)	1,398	59.6	16.1	1.43	.40	18	59.8	14.9	1.38
Japan	7,610	67.3	16.8	.29	.04	0.5	66.8	16.1	.28
Korea	4,522	60.2	15.3	.48	.15	5	61.0	15.3	.49
Netherlands	5,025	65.8	16.9	.86	.50	16	63.7	16.1	.83
Nigeria	804	40.8	14.3	1.08	.31	88	42.2	13.8	1.16
Norway	1,420	59.8	15.8	.52	.02	1	59.3	14.7	.48
Papua New Guinea	2,193	54.5	12.5	.57	.11	13	55.3	11.4	.56
Philippines	10,871	38.2	15.4	.67	.48	87	39.7	14.4	.65
Poland	4,520	60.4	17.3	.74	.34	14	59.5	15.9	.66
Singapore	4,430	54.9	16.2	.93	na	32*	56.4	15.9	.96
Sweden (Grade 7)	1,557	57.7	15.6	.76	.09	7	56.0	15.1	.70
Sweden (Grade 8)	1,461	61.4	16.3	.74	.08	2	60.3	16.0	.81
Thailand	3,780	55.1	13.6	.74	.24	26	56.7	12.6	.74
U.S.A.	2,519	54.8	16.7	.91	.29	30	-	-	-
Zimbabwe	2,648	41.3	14.4	.62	.13	80	42.8	11.6	.66

* classes and not schools

The ROH index was explained above for Table 3.1 for Population 1. At Population 2 level the Nordic countries and Japan have negligible differences between classes and schools.

Science Achievement

In Ghana (ROH = .48), Italy (Grade 9) (ROH = .40), the Netherlands (ROH = .50) and the Philippines (ROH = .48), the class or school which the child attends has a great deal to do with the science test scores he or she achieves. In Italy (Grade 9) and the Netherlands this reflects a differentiated school system with several school types for "different ability levels" of students. Italy and the Netherlands may have clearly a different philosophy about grouping students than do the Nordic countries or it may be a result of the administrative way in which the system is organized. In Ghana and the Philippines, one wonders if this reflects an urban-rural dimension with large differences in how the schools/classes are equipped and staffed.

A plot was made to examine the relationships associated with the between (B) and within (W) school standard deviations across populations, relating the B for Population 1 to the B for Population 2 and similarly for the Ws, but are not presented because of space constraints. The between standard deviations showed a strong positive relation over grades (r = .72) for the 16 country combinations with data at both time points. [Sweden (3) with (7) and Sweden (4) with (8) were paired.] Generally, the Nordic countries and Japan had low between components in both populations and Nigeria and the Philippines were outliers on the high side. Removal of the outlying points increased the correlation from .72 to .77. The between standard deviations are very similar for the two populations. The plot of the W standard deviations, on the other hand, showed no relationship whatsoever.

Substantively, the above findings imply that, compared with other societies, the Nordic countries and Japan have strong societal factors which homogenize science achievement across classes and schools, but which leave within school variation similar to that over most of the rest of the world. To be true of both populations, these must be educational rather than background effects.

On the other hand, Nigeria and the Philippines are atypical in having larger between school, together with smaller within school variations, than all other countries. This pattern must involve the processes by which student are allocated to classes and schools as well as educational effects. This pattern must be partially caused by processes which constrain which students are allowed to attend specific schools and classes.

The "% of schools below 49.4" column presents the percentage of schools or classes in the sample in each country scoring lower than the lowest scoring class in Hungary which was the highest scoring country. It is interesting to note that the United States has a higher percentage of classes than Thailand, scoring lower than the lowest class in Hungary!

Table 3.4 presents the subscore results for countries. The countries have been classified into groups. The range of scores used for the groups is given at the foot of the table.

From a perusal of the table the following statements can be made:

- Hungary and Japan tend to score consistently well.
- China and The Netherlands score particularly well in physics, chemistry and comprehension, but poorly in biology.
- Ghana, Italy (Grade 8), Nigeria, the Philippines and Zimbabwe have consistently low scores compared with other countries.

TABLE 3.4 Classification of countries by score group for total and sub-scores* (Population 2)

Score range	Core+2rotated	Biology (23 items)	Chemistry (15 items)	Earth Science (9 items)	Physics (23 items)	Information (19 items)	Comprehension (24 items)	Application (27 items)
more than 65%	HUN JAP	HUN JAP	HUN	AUS CAE HUN ISR IT9 JAP KOR NET SW8	CHN HUN JAP NET	HUN	CHN HUN JAP NET	HUN JAP
60-65%	CAE CHN FIN KOR NET SW8	CAE CAF FIN IT9 NET POL THA	CHN NET	CAF CHN FIN HKO NOR POL SW7	AUS CAE CAR FIN IT9 KOR NOR SW8	CAE NET NOR	CAE FIN IT9 KOR POL SW8 THA	CAE CAF IT9 KOR NET SW8
55-59%	AUS CAF ENG HKO ISR IT9 PNG NOR POL SIN SW7 THA	AUS ISR KOR NOR PNG SIN SW8	JAP FIN ISR SW8 POL NOR	ENG IT8 PNG SIN THA	ENG HKO ISR PNG POL SIN SW7 THA	AUS CHN FIN ISR IT9 JAP KOR SIN SW8 PNG POL	AUS CAF ENG HKO ISR NOR PNG SIN SW7	AUS CHN ENG HKO ISR NOR POL SW7 THA
50-54%	IT8	CHN ENG HKO IT8 SW7	CAE KOR SW7 IT9 AUS		IT8	CAF ENG HKO IT8 SW7 THA	IT8	IT8 PNG SIN
less than 50%	GHA NIG PHI ZIM	GHA NIG PHI ZIM	CAF ENG GHA HKO IT8 NIG PHI PNG SIN THA ZIM	GHA NIG PHI ZIM	GHA NIG PHI ZIM	NIG ZIM GHA PHI	GHA NIG PHI ZIM	GHA NIG PHI ZIM
	Median 55.2%	Median 58.5%	Median 50.5%	Median 63.7%	Median 59.5%	Median 55.9%	Median 59.8%	Median 58.6%

AUS - Australia; CAE - Canada (English speaking); CAF - Canada (French speaking); CHN - China; ENG - England; FIN - Finland; GHA - Ghana; HKO - Hong Kong; HUN - Hungary; ISR - Israel; IT8/IT9 - Italy; JAP - Japan; KOR - Korea; NET - Netherlands; NIG - Nigeria; NOR - Norway; PNG - Papua New Guinea; PHI - Philippines; POL - Poland; SIN - Singapore; SW7/SW8 - Sweden; THA - Thailand; ZIM - Zimbabwe.

* The actual percentage subscores from which this table is derived are given in Appendices F.3 and F.4.

Science Achievement

Population 3

Before reporting the scores for Population 3 it is desirable to refer back to Table 1.2 in Chapter 1 of this volume. There are certain 'facts' to be borne in mind when examining the scores of Population 3.

It will be recalled that Population 3 Science was defined as those students studying science for examination purposes in order to study science further at the higher education level. This is the equivalent of what is referred to as 'specializing' in science. This is not an easy matter to define in each of the different systems. For example, in Finland 41 percent of an age group is enrolled in academic schools; another 18 percent are enrolled in vocational schools and four percent in evening schools and only those in academic schools were tested. However, in the academic schools, biology is a compulsory subject. In Hungary, on the other hand, 18 percent of an age group is enrolled in academic schools and 22 percent in vocational schools. Because the vocational schools do not normally lead to higher education these were excluded. All students in the academic school must study general science but some take extra enrichment courses in biology or chemistry or physics. Hungary classified those in enrichment courses as the 'specialists' and for biology, for example, this was only three percent of an age group. This is an extreme case of the problem of population definition but points to the difficulty of comparisons at this population level.

The following should be borne in mind when making comparisons:

a) Percentage of an age group studying
   - the overall percentage of an age group enrolled in all schools at this level (ranging from one percent in Ghana and Papua New Guinea to 89 percent in Japan);
   - the percentage of an age group defined in the populations for biology, chemistry, and physics: for biology it ranges from three percent in Hungary and Singapore to 41 percent in Finland; for chemistry it ranges from one percent in Ghana, Hungary, and Italy to 37 percent in Canada (Fr.) and Korea; for physics, it ranges from one percent in Ghana and the United States (a second year in physics) to 35 percent in Canada (Fr.).

b) Average age
   There was a 23 month difference in average age between the groups ranging from 17 years and 2 months in Canada (Fr.) to 19 years and two months in Hong Kong (Form 7 or Grade 13).

c) Grade level
   Again there is a two year difference in the grade level tested. Although for most systems, it was Grade 12, it was Grade 11 for Canada (Fr.) Québec and Grade 13 for some schools in Canada (Eng.), England, Ghana, Hong Kong (Form 7), (Italy, in the main) and Singapore (the third year of the Pre-university Centres).

d) Average number of subjects studied at this level
Whereas in most countries the students study five, six, or seven subjects there are two countries (England and Ghana) where students only study an average of three subjects and five countries where nine or more subjects are studied (Finland, Hungary, Korea, Poland, and Sweden). In these countries, it is more likely that it is a system of general education rather than highly specialized in science.

e) Degree of overlap between the sub-populations taking biology, chemistry and physics
In some countries students are probably studying only one of the branches of science at Population 3 level. This is the case in the United States. In other countries students are taking one, two, or even three science subjects; a common combination is chemistry and physics. In Finland, all students study biology and only parts of the age group study the other two branches. In Hungary, all study science and some only take enrichment courses in one or two of the subjects.

Finally, before proceeding to the scores it should be indicated that very few of the research centers in the study were able to obtain estimates, from the statistics divisions of their ministries of education, of the number of students in the country studying biology, chemistry or physics at that level. However, the ministries did, in general, know how many students were enrolled altogether in Grades 11, 12 or 13. It is the relationship between the number of students in a school or stratum studying biology or chemistry or physics and all students in school at that level which was used for weighting to correct for disproportional drop-out between strata.

Biology. Table 3.5 presents the scores for biology.
It will be seen that it is Singapore, Hong Kong, and England which have the highest total scores. In Singapore and in Hong Kong students study an average of four to six subjects while in England it is three subjects. However, Hungary with an average of more than nine subjects studied is a close runner-up. All of these countries, however, have only a small percentage of an age group studying the subject.

Canada (Fr.) which consists mostly of Grade 11 students from Québec taking their second course in biology, together with Korea, Italy, and Thailand have low scores. Korea has 38 percent of an age group studying biology. Finland, with 41 percent of an age group and Israel with 20 percent are near the mean score.

The ROH values for Hungary, Israel and Japan are high. The remarkable increase in the size of ROH in Japan from Population 2 to 3 is probably due to the large increase in the number of private schools which emerge at this level of schooling and to the entrance examination for Grade 10 where a selection process into different upper secondary schools occurs.

The lowest scoring class in Singapore had a mean of 61.6 percentage points. It is interesting to note the high percentage of schools and classes in other countries scoring lower than the lowest class in Singapore.

TABLE 3.5 Means and standard deviations in biology for Population 3 (in percent)

Country	N	Total (N=39 items)			ROH	% schools below 61.6	Information (N=13 items)		Comprehension (N=16 items)		Application (N=10 items)	
		Mean	sd	se			Mean	sd	Mean	sd	Mean	sd
Australia	1,631	53.4	13.3	.50	.12	90	52.8	17.2	51.4	16.8	57.3	17.2
Canada (Eng.)	3,254	48.7	14.3	.63	.25	96	47.2	17.6	46.8	17.7	53.6	18.2
Canada (Fr.)	249	42.3	12.7	1.14	.17	93	38.9	15.5	39.6	17.0	51.1	17.0
England	884	68.1	11.9	.74	.21	14	70.3	13.7	66.3	15.6	68.2	16.4
Finland	1,652	51.0	11.5	.51	.04	95	52.2	15.6	48.4	15.3	53.7	16.1
Ghana	210	63.6	11.5	1.20	.16	35	69.1	16.1	60.2	13.9	61.8	16.0
Hong Kong (F.6)	2,975	61.0	12.1	.83	.19	43	64.6	16.7	59.1	14.3	59.4	16.5
Hong Kong (F.7)	1,559	70.2	13.0	.98	.20	7	75.3	16.6	67.6	14.9	67.7	16.6
Hungary	301	65.6	12.3	1.13	.38	37	67.4	14.7	64.0	16.1	66.0	15.4
Israel	879	56.9	19.6	1.98	.43	63	56.9	22.5	53.9	21.6	61.8	21.3
Italy	147	46.7	14.1	3.49	.29	100	50.0	16.5	41.0	16.3	51.6	19.3
Japan	1,212	51.3	15.4	1.70	.44	82	46.7	17.5	54.6	19.1	52.1	18.5
Korea	3,319	44.5	14.1	.56	.27	99	43.8	16.9	43.9	18.6	46.5	17.2
Norway	276	59.9	13.9	.96	.06	63	61.4	16.5	55.7	17.4	64.6	17.7
Poland	764	59.1	12.3	.90	.32	69	66.9	15.4	52.3	14.7	59.9	18.0
Singapore	902	70.9	11.3	1.44	.11	0	71.1	14.8	71.1	14.4	70.3	15.5
Sweden	619	60.5	14.0	.77	.09	48	56.9	17.9	61.3	16.8	63.7	16.2
Thailand	1,171	47.2	12.7	.83	.32	94	43.7	16.3	45.7	16.0	54.3	17.6

The scores on information, comprehension and application type items tend to mirror the total score differences among countries. Perhaps of most note is Japan's relatively low performance in biology at this level. At the same time, Japan has 63 percent of an age group in Population 3 but only 12 percent of those in school study biology. It should be noted, however, that girls dominate the biology classrooms in Japan (see Table 2.9) while the boys are concentrated in chemistry and physics classrooms.

Chemistry. Table 3.6 presents the scores for chemistry.

The total mean score for all countries is 52.5. There is more variation in score than for biology. Again, Hong Kong (Form 7), England and Singapore and Hong Kong (Form 6) have high scores. Although England and Singapore have only five percent of an age group studying chemistry, in Hong Kong Form 7 has 12 percent but Form 6 has 20 percent for chemistry. The countries with scores equal to or less than 40 percentage points are Canada (Fr.) (37 percent of an age group studying chemistry), Korea (37 percent of an age group studying chemistry), Thailand (seven percent of an age group studying chemistry), and Finland (16 percent of an age group studying chemistry).

Physics. Table 3.7 presents the scores for physics.

The total mean score for all countries is 49.9. Again Hong Kong (Form 7) with 12 percent of an age group studying physics has by far the highest score. It also has the highest scores in information, comprehension, and application. England (six percent of an age group) and Hong Kong (Form 6) (20 percent) followed by Hungary (four percent) and Japan (11 percent) come next. The low scores are from Canada (Fr.) (35 percent of and age group and mostly Grade 11), Italy, and Thailand.

What then can be said about the overall achievement results at Population 3 level?

- Hong Kong (Form 6) which is 12th Grade has higher scores than most other countries. This indicates what is possible. One might then query the necessity of a 13th Grade in some countries. The 13th Grade (Form 7) in Hong Kong has very high achievement compared with other countries and is consistently good across the three science subjects and across each of the three cognitive levels.
- England and Hungary score relatively highly but both involve élite groups of students.
- Japan is relatively weak in biology compared with its performance in chemistry and physics.
- Canada (Fr.) which is comprised primarily of Grade 11 students in Québec has a low performance compared with other countries and also, a much lower level of achievement relative to Canada (Eng.) which is one grade, and sometimes two grades, older than the French-speaking Canadians.

TABLE 3.6 Means and standard deviations in chemistry for Population 3 (in percent)

Country	N	Total (N=39 items) Mean	Total sd	se	ROH	% schools below 38.1	Information (N=9 items) Mean	Information sd	Comprehension (N=16 items) Mean	Comprehension sd	Application (N=14 items) Mean	Application sd
Australia	1,177	53.5	16.7	.81	.19	4	48.8	20.2	56.9	19.7	52.5	18.3
Canada (Eng.)	2,923	43.1	14.6	.62	.23	29	35.8	17.2	47.6	18.7	42.6	16.7
Canada (Fr.)	1,187	32.1	11.3	.60	.13	87	24.1	14.4	35.6	15.8	33.3	13.9
England	892	73.1	15.0	.82	.20	0	71.2	19.5	76.3	17.8	70.7	16.4
Finland	971	40.5	13.3	.87	.13	24	34.4	16.7	44.4	17.5	40.7	15.7
Ghana	351	63.7	16.6	2.07	.25	5	69.0	18.6	67.4	18.8	56.1	20.1
Hong Kong (F.6)	5,952	66.9	14.7	1.20	.31	1	64.0	19.2	74.6	16.2	59.8	17.7
Hong Kong (F.7)	3,599	78.5	14.6	1.15	.24	0	74.5	19.0	84.6	15.2	74.1	17.3
Hungary	143	53.4	16.7	2.19	.46	13	51.7	19.8	57.3	20.5	50.0	17.7
Israel	243	50.3	19.5	4.13	.43	19	43.7	20.9	57.9	23.0	45.8	20.1
Italy	217	42.9	22.4	4.76	.63	38	43.4	23.2	46.4	26.4	38.6	23.2
Japan	1,468	57.8	20.2	2.66	.64	14	46.7	22.4	65.0	22.4	56.6	22.2
Korea	2,979	33.9	13.8	.56	.27	65	29.9	18.4	34.6	18.0	35.7	15.0
Norway	283	47.9	15.7	1.16	.12	4	39.3	18.9	54.3	18.9	46.2	18.6
Poland	765	49.5	15.8	1.34	.42	15	48.5	19.8	56.6	17.8	42.0	18.3
Singapore	945	68.7	15.7	3.15	.27	0	67.0	20.1	73.5	17.4	64.3	18.3
Sweden	1,172	47.3	15.0	.71	.17	12	44.6	18.2	50.2	18.7	45.6	17.0
Thailand	1,168	38.8	15.7	1.17	.43	57	37.1	18.9	39.7	20.1	38.8	16.7

TABLE 3.7 Means and standard deviations in physics for Population 3 (in percent)

Country	N	Total (N=38 items)				% schools below 57.1	Information (N=9 items)		Comprehension (N=20 items)		Application (N=9 items)	
		Mean	sd	se	ROH		Mean	sd	Mean	sd	Mean	sd
Australia	1,073	53.6	14.6	.68	.18	71	48.9	20.3	51.6	15.9	62.6	18.5
Canada (Eng.)	2,766	47.0[a]	14.5	.67	.26	91	39.6	20.2	47.0[c]	16.3	54.2	19.0
Canada (Fr.)	944	30.2[b]	10.7	.57	.17	100	26.7	15.8	27.4[d]	12.4	39.8	17.9
England	917	64.2	13.3	.60	.14	19	56.5	19.2	63.5	15.0	73.5	16.1
Finland	810	44.7	13.5	.90	.10	98	37.8	18.9	42.5	15.5	56.7	17.8
Ghana	291	54.3	13.5	1.43	.14	75	53.1	18.7	54.5	15.4	55.0	17.8
Hong Kong (F.6)	5,906	64.4	13.0	1.08	.34	10	55.2	20.3	63.5	14.5	75.8	15.6
Hong Kong (F.7)	3,690	73.7	12.6	1.10	.27	0	67.4	18.7	73.1	14.5	81.3	14.1
Hungary	398	62.7	15.3	1.50	.43	41	54.6	19.5	61.4	17.7	73.6	16.2
Israel	472	54.4	17.0	1.42	.17	67	46.5	21.8	52.7	17.8	66.2	21.5
Italy	1,766	34.2	13.4	.89	.41	99	26.7	17.2	29.6	14.6	51.9	19.9
Japan	1,187	61.8	16.1	1.85	.44	42	53.2	21.2	60.6	18.3	73.1	17.6
Korea	981	43.3	17.2	1.10	.27	88	39.2	22.0	40.6	19.5	53.5	19.6
Norway	443	56.5	15.5	1.06	.12	51	58.1	21.2	52.5	16.3	63.6	21.1
Poland	1,716	56.4	16.5	1.33	.46	54	58.4	21.7	52.4	18.5	63.1	18.1
Singapore	1,071	59.2	12.8	1.31	.07	38	50.7	18.4	59.1	14.8	68.1	17.0
Sweden	1,156	51.4	14.5	.63	.12	79	48.2	19.8	48.5	16.1	61.0	18.4
Thailand	1,168	35.9	16.9	1.31	.47	93	33.6	20.8	33.0	18.8	44.3	20.5

[a] 37 items because one item (P3P07) was presented with different formulation.
[b] 37 items because one item (P3P09) was presented in an incorrect form.
[c] 19 items because the differently formulated item belongs to subscale "Comprehension".
[d] 19 items because the incorrectly formulated item belongs to subscale "Comprehension".

Science Achievement

The United States Scores. It will have been noticed that the scores for the United States are missing from the three sets of scores thus far presented for Population 3. As pointed out in Chapter 1, the United States, did not administer the core test to Population 3 students and only administered 25 of the 30 item biology test, 25 items of the 30 item chemistry test and 26 items of the 30 item physics test (see Appendix B Part 2).

Thus, a special set of analyses was computed using only the items taken by the United States. Since the United States made a special selection of items from the tests, it is assumed that those responsible selected the items which were thought to be most relevant to the United States. If this were so, then the United States should have an advantage over other countries on both item scores and total scores. However, the restricted number of items does mean that the content coverage is not as large as would be desired.

Table 3.8 presents the three sets of scores. The data are presented for purposes of comparison and are limited only to those items taken by the United States. The countries other than the United States can be seen to have similar relative performance to that presented in Tables 3.5 to 3.7.

In these tables, the United States has the lowest scores in biology, the third lowest in chemistry, and the seventh lowest in physics; in short, not a very good performance for advanced science students. In biology and physics all United States schools scored lower than the lowest school in the highest scoring country. The ROH values for the United States are .40, .49, and .39 for biology, chemistry,

TABLE 3.8 *Biology, chemistry and physics achievement for Population 3 as measured on the reduced number of items taken by the United States*

Country	Biology (25 items)			Chemistry (25 items)			Physics (26 items)		
	Mean	sd	se	Mean	sd	se	Mean	sd	se
Australia	46.6	14.2	.52	49.1	19.5	.91	48.7	15.2	.70
Canada (Eng.)	43.7	14.4	.60	39.6	16.8	.70	41.7	14.6	.65
Canada (Fr.)	39.3	13.4	1.32	27.9	12.3	.64	26.7	10.6	.53
England	62.4	13.6	.80	69.3	17.9	.94	58.4	15.1	.68
Finland	50.2	13.1	.56	35.9	15.3	.96	37.9	14.1	.96
Ghana	57.6	13.1	1.46	60.2	18.3	1.97	50.5	13.3	1.29
Hong Kong (Form 6)	55.2	14.5	1.00	68.2	17.7	1.37	61.2	14.5	1.12
Hong Kong (Form 7)	65.0	16.0	1.13	78.5	17.6	1.28	70.8	14.2	1.16
Hungary	59.9	14.0	1.21	50.2	19.7	2.43	58.7	17.6	1.74
Israel	51.6	22.5	2.16	45.1	24.8	5.31	54.5	18.7	1.50
Italy	42.3	15.0	3.59	38.2	24.0	4.93	29.2	13.6	.85
Japan	45.5	15.4	1.56	55.5	22.9	2.98	58.5	17.8	2.04
Korea	40.1	14.1	.51	30.9	14.8	.54	39.8	16.9	.99
Norway	55.4	15.1	1.09	44.3	18.4	1.32	54.1	15.9	1.05
Poland	56.3	13.2	.96	46.5	17.4	1.45	53.2	17.1	1.38
Singapore	66.1	13.2	1.54	65.7	17.7	3.39	55.0	13.3	1.26
Sweden	53.4	15.8	.84	43.9	17.8	.84	46.0	15.0	.59
Thailand	42.6	14.2	.91	35.8	16.5	1.15	34.3	16.7	1.26
U.S.A.	38.1	15.5	1.62	37.7	18.2	2.60	45.3	15.9	2.07

and physics respectively. These high ROH values indicate a large disparity between schools, particularly for a rich country priding itself on equality of educational opportunity.

There is clearly something sadly wrong with science education in the United States and this should be of immediate concern to all United States politicians and educators. It is sometimes said that the achievement in secondary schools in the United States does not matter because the deficiencies in secondary school are recouped at the college level. There would seem to be little evidence for such a claim. Only an international testing program of college undergraduates could establish the truth or falsity of the conjecture.

The Élite of Population 3. The average scores for those studying biology, chemistry, and physics have been seen. But what of the scores of the élite in these subjects? The scores for Ghana are dropped from these calculations because Ghana had only .2 or .6 percent of an age group studying at this level and to have taken this percent as a basis on which to compare would have resulted in too few students in other countries for any stable estimate. Therefore, the lowest percentage of an age group studying biology was to be found in Hungary and Singapore. This was three percent and therefore the top three percent of an age group was compared across countries. It can be seen that in Australia 18 percent of an age group was studying biology. Therefore the top sixth of these 18 percent of students in terms of achievement scores was taken, the top three quarters of those in England and so on. In chemistry, the lowest percentage for which stable estimates could be calculated was five percent. Ghana, Hungary and Italy were omitted because the N would be too small for stable estimates. For physics it was four percent (Hungary).

Table 3.9 presents the results.

The assumption is that, for example, in biology the top three percent in Singapore and Hungary are the best three percent of the age group. It is just possible (but not likely) that there may be some not in school who would achieve better than those in school. But, given that many systems ask the question "if a larger percentage of an age group is allowed through to 12th Grade, will the performance of the élite be affected?", it was considered desirable to examine the question using this form of analysis.

It can be seen in biology that the top three percent in Hong Kong (7) and Israel do particularly well whereas Thailand performs relatively poorly. In chemistry, it is the élite in both forms in Hong Kong and Japan which perform well and Canada (Fr.), Norway, and Thailand which perform poorly. For physics, the élites of both forms in Hong Kong and Japan perform well and Canada (Fr.), Italy, and Thailand perform poorly.

Science Achievement 71

TABLE 3.9 Means and standard deviations for upper three, five and four percent of an age group for Population 3 (in percent) for biology, chemistry and physics (the élite students)

Country	Upper 3% Biology			Upper 5% Chemistry			Upper 4% Physics		
	N	Mean	sd	N	Mean	sd	N	Mean	sd
Australia	272	72.7	5.6	490	69.6	9.6	390	69.1	7.8
Canada (Eng.)	349	73.0	5.1	585	65.4	7.7	615	67.0	7.9
Canada (Fr.)	107	52.4	7.7	160	51.7	6.2	108	49.7	5.7
England	663	73.5	8.2	892	73.1	15.1	611	71.5	9.0
Finland	121	78.0	3.6	303	56.5	9.0	231	60.7	8.7
Hong Kong (F.6)	744	75.4	4.8	2,480	84.2	4.5	1,969	82.3	4.9
Hong Kong (F.7)	668	80.6	4.8	1,500	89.7	4.0	1,230	87.1	4.6
Hungary	301	65.6	12.4	*	*	*	398	62.6	15.2
Israel	132	79.9	4.1	152	62.8	8.9	189	69.9	7.3
Italy	32	66.3	6.6	*	*	*	372	53.1	8.2
Japan	303	70.8	6.8	459	81.0	7.5	432	78.4	7.5
Korea	158	71.7	5.4	596	59.9	9.1	327	65.0	9.7
Norway	207	66.0	10.2	236	52.0	13.1	177	71.8	8.8
Poland	255	72.1	5.7	425	61.0	9.7	763	71.3	9.0
Singapore	902	70.9	11.3	945	68.7	15.7	612	68.1	7.7
Sweden	371	69.5	7.6	391	62.7	9.5	356	68.1	8.3
Thailand	502	59.2	7.6	834	45.6	13.2	667	46.4	14.8

NOTE: Ghana was dropped because of the very low proportion of an age group in school.
* Not enough data for comparison.

Population 3 Scores, Percentage of an Age Group in School, Grade, Age, and Subjects Studied

Several important questions may be asked about the élite scores and Population 3 scores in general:

1. Is the performance of those studying biology, chemistry and physics (and the élite in those groups) related to the overall percentage of an age group in school?
2. Is the performance of all those studying biology, chemistry and physics lower the greater the percentage of an age group enrolled in those subjects at that level?
3. Is the performance of the élite lower the greater the percentage of an age group studying the subject in school?
4. Do countries where more subjects are studied have a lower performance in the science subjects both for the élite and for all students?
5. Do countries where the students are older perform better than countries where the students are younger?

The following between-country correlations (Table 3.10) provided some evidence with which to answer the questions. The general trend is of interest.

TABLE 3.10 Selected between-country correlations at Population 3 level

	All studying			Élite studying		
	Biology	Chemistry	Physics	Biology	Chemistry	Physics
Percent of all studying in school	-.59*	-.47*	-.34	-.11	-.13	-.12
Percent studying biology, chemistry, physics	-.54*	-.54*	-.48*	.36	-.10	-.27
Average No. of subjects studied	-.31	-.55*	-.29	.06	-.37	-.13
Average age	.46	.28	.44	.31	.18	.42
Grade	.68*	.74*	.74*	.65*	.57*	.59*

* indicates that the correlation is significant at least at the 95% level.

Q1: The surface answer to question 1 is 'Yes' for those studying biology (r=-.59), chemistry (r=-.47) and physics (r=-.34) but 'No' for the élite. However, the actual answer is 'No'. A multiple regression analysis involving both science and overall percentages clearly indicates that the reason for the correlations between overall enrollment percentages and achievement is their correlation with a third set of variables: the percentages of an age group studying science. The surface finding is therefore spurious. That is, no evidence is found for true effects of overall enrollment percentages, once the percentages studying science are taken into account.

Q2: The answer to question 2 is 'Yes'. That is the greater the proportion of an age group enrolled in biology, chemistry, or physics the lower the average score. This could be due to the increase in the range of ability amongst the students. However, there is no significant influence on the élites' performance since they are the exceptional amongst the students. Again, this conclusion is confirmed by more refined analyses. Regression analyses involving both science subject enrollment percentages and overall percentages clearly indicate negative effects for the former, independent of the latter.

Q3: The answer to question 3 is that none of the correlations is significant indicating that there is no evidence of a relationship between the percentage of an age group in school studying a subject and the performance of the élite. However, the signs should be noted. For chemistry and physics the signs are negative but for biology the sign is positive and the correlation is +.36 indicating that the greater the percentage of an age group enrolled in biology, the performance of the élite is higher, but not significantly so. Careful scrutiny using various combinations of the percentages confirms that there is no statistical evidence of any significant relationship, either to overall or science enrollment percentages.

Q4: The answer to question 4 is 'No' except for those studying chemistry. Although only one of the relationships is significant, the negative trend does point to the fact that students who are enrolled for more subjects are less likely to be science oriented. This is because some *de facto* specialization does occur in all countries.

Q5: The surface answer to question 5 would appear to be 'Yes' because all correlations are positive. However, age, as would be expected, is highly correlated with grade in this sample of countries. Grade is an alternative explanation because it is a more direct surrogate for total exposure to science instruction than age. In most countries, the grade for Population 3 is constant at 12. There is only one country lower, and only a few higher. Even given this low level of variation, the grade correlation with achievement is higher than that of age. When a regression analysis is performed with both grade and average age, the coefficient for age is cut by a factor of seven and is still non-significant - and that for grade is positive and highly significant. Hence, the higher the grade level, the higher the achievement.

What are the implications of these findings for educational planners? If participation rates in science subjects are increased, then it is to be expected that the mean achievement will be lower for all of those studying each of the science subjects. On the other hand, it may be concluded that the mean achievement of the élite does not suffer as a result of increased participation rates. With the exception of chemistry there is no significant relationship between achievement and the number of subjects studied. Hence, it is suggested that in those countries where it is common for students to study only three or four subjects, careful experimentation could be undertaken to increase the number of subjects to five or six and hence gain a good balance between arts and science subjects. Indeed, such a change is now being considered in England. There is evidence that more learning occurs in those systems with more grades. How systems decide on the break between secondary and higher education must be tied to tradition. However, Canada (Fr.) where Grade 11 is the last grade of secondary school may well wish to consider having a Grade 12.

Relative Increase of Achievement over Populations and Sex Differences

From the data presented in Chapter 2 and this chapter some relationships within and across populations have been examined.

Relative Increase of Achievement from Populations 1 to 2 and 2 to 3

Fourteen countries tested at all three population levels. These were: Australia, Canada (Eng.), Canada (Fr.), England, Finland, Hong Kong, Hungary, Israel, Italy, Japan, Korea, Norway, Poland, and Singapore. There were 13 items which were the same anchor items on the core tests for Populations 1 and 2. Likewise, 15 items were common to the core tests for Populations 2 and 3.

Using these anchor items for the 14 countries, means and standard deviations were calculated for each population level. Using Population 2 overall standard deviations, standard scores were calculated for the overall means of Population 1 and 3. These were, in fact, -1.2 and +1.2 respectively. Using the overall standard deviation of all Population 1 core scores and taking -1.2 to be the zero point for Population 1 scores, standard scores were then computed for each Population 1 national score. The same exercise was repeated for the Population 3 scores. For Population 3 two sets of calculations were carried out on the core test scores: one for those taking the biology, chemistry and physics tests, and a second for those taking the 3N (non-specialist) tests. The resultant standard scores were then multiplied by the ratio of the standard deviations (Population 1/Population 2) for the Population 1 scores similarly (Population 3/Population 2) for the Population 3 scores, thus allowing all national scores to be placed on one overall standard score scale.

Table 3.11 presents the mean scale scores for each country and Figure 3.1 presents these scores in diagrammatic form.

TABLE 3.11 Mean scale scores across Populations (1, 2, 3S, 3N)

Country	Population 1	Population 2	Population 3S (Specialists)	Population 3N (Non-specialists)
Australia	-1.54	-.04	1.39	.33
Canada (Eng.)	-1.30	.12	1.13	-.30
Canada (Fr.)	-1.09	.01	.46	-.33
China (Pop.2 only)		-.08		
England	-1.85	-.27	2.11	.92
Finland	-.86	.11	1.26	
Ghana (Pop. 2 & 3 only)		-.93	1.42	
Hong Kong (Form 6)	-2.00	-.35	2.21	1.94
Hungary	-1.10	.78	2.13	1.65
Israel	-1.80	.12	1.39	.09
Italy (Grade 8 for Pop. 2)	-1.40	-.49	.75	.10
Italy (Grade 9 for Pop. 2)	-1.40	-.02	.75	.10
Japan	-.83	.47	1.77	.88
Korea	-.85	.01	.57	
Netherlands (Pop. 2 only)		.37		
Nigeria (Pop. 1 & 2 only)	-2.90	-1.23		
Norway	-1.59	.01	1.48	.27
Papua New Guinea (Pop. 2 & 3 only)		-.35		.05
Philippines (Pop. 1 & 2 only)	-2.46	-1.40		
Poland	-1.79	.02	1.47	
Singapore	-1.98	-.33	1.85	.45
Sweden (Grade 3 for Pop. 1 & Grade 7 for Pop. 2)	-1.55	.15	1.82	.61
Sweden (Grade 4 for Pop. 1 & Grade 8 for Pop. 2)	-1.05	.09	1.82	
Thailand		-.31	.70	-.62
U.S.A. (Pop. 1 & 2 only)	-1.46	-.33		
Zimbabwe (Pop. 2 only)		-1.20		

The average international year growth over four years from Population 1 to Population 2 is .35. Israel and Hungary have high yearly growths (.48 and .47 respectively) but Korea and Finland have low yearly growths (.22 and .24 respectively). This is not surprising given the high relative scores of these two countries at Population 1.

The average yearly growth from Population 2 to Population 3 is considered in two parts: the first is for those taking the biology, chemistry or physics tests (called 3S) and the second is for Population 3N (the non-specialists).

The average yearly growth from Population 2 to Population 3S is .42, and to 3N is .12. For Population 3S (i.e. those studying science) England, Ghana, Hong

Kong, and Singapore have high values (.68, .92, .65, and .62 respectively). This, again, reflects the selection of good students into the courses and the high degree of specialization. Canada (Fr.) and Korea have low values (.13 and .16 respectively) reflecting the fact that the vast majority of students tested in Canada (Fr.) were from Grade 11 and that a high percentage of an age group in Korea were in 3S. For the international year growth for Population 3N, England has a high value of .34. This presumably reflects the fact that all, or nearly all students in school, study science to the age of 16 years. Furthermore, although those continuing in the arts track to the 'A' level have none or little further science, they still remember a good deal of science. However, there are certain countries where the 3N Population has a negative yearly value. These are Canada (Eng.) -.04, Canada (Fr.) -.10, Israel -.01, and Thailand -.09.

It is of interest to note that, at Population 2 level, Ghana, Nigeria, the Philippines, and Zimbabwe have lower scores (despite the fact that they are older) than the 10-year-olds in Finland, Japan, and Korea. If one assumes that the rate of development in science learning is the same below ten years of age as between ten and fourteen years old and if one applies the average yearly increase (.35 from Population 1 to Population 2) and extrapolates downwards then it can be tentatively said that, when compared with Japan, the Philippines Grade 9 has about the same level of achievement as Grade 3 in Japan. Nigeria (Grade 10) and Zimbabwe (Grade 9) have the same achievement as Grade 4 in Japan. Indeed, one must ask oneself if the children in some 'developing' countries (Nigeria, Philippines and Zimbabwe) are taking 1.5 to 2.0 more time to reach the same level of achievement as some 'developed' countries.

Sex Differences

Table 3.12 presents the standard score (boys' minus girls' scores divided by average total standard deviation for all countries) differences for boys and girls for each population.

The average sex differences in overall achievement in favour of boys increase from Population 1 to Population 2 level. Japan, Hungary, and Sweden (Grade 3) all have low sex differences at Population 1. All have high average total scores. The Philippines also has a low standard score sex difference but it also has a low mean score.

At Population 2, China and the Netherlands have very high sex differences. Hungary, the highest scoring country, and the Philippines, a low scoring country, have the least standard score difference.

At Population 3 (based on separate tests of biology, chemistry, and physics plus the appropriate items from the core test), however, there is a mixed picture. The highest overall sex differences are in physics and the lowest are in biology. Within each subject area there is also wide variation. In Israel, girls score better than boys in biology and chemistry. In biology, there are no differences or very small differences between boys and girls in Australia, Canada (French), and Ghana. In chemistry, this is true in Ghana and Sweden. In physics, it is true in England and Singapore.

Some countries have dramatic differences between subjects (e.g. Australia, Finland, Hungary, Italy, and Poland). On the other hand, other countries have very similar differences among the subjects (e.g. Canada (English), Japan, Korea, Norway, Singapore, and the United States).

FIG. 3.1   Standard score differences across Populations (1, 2, 3S, 3N)

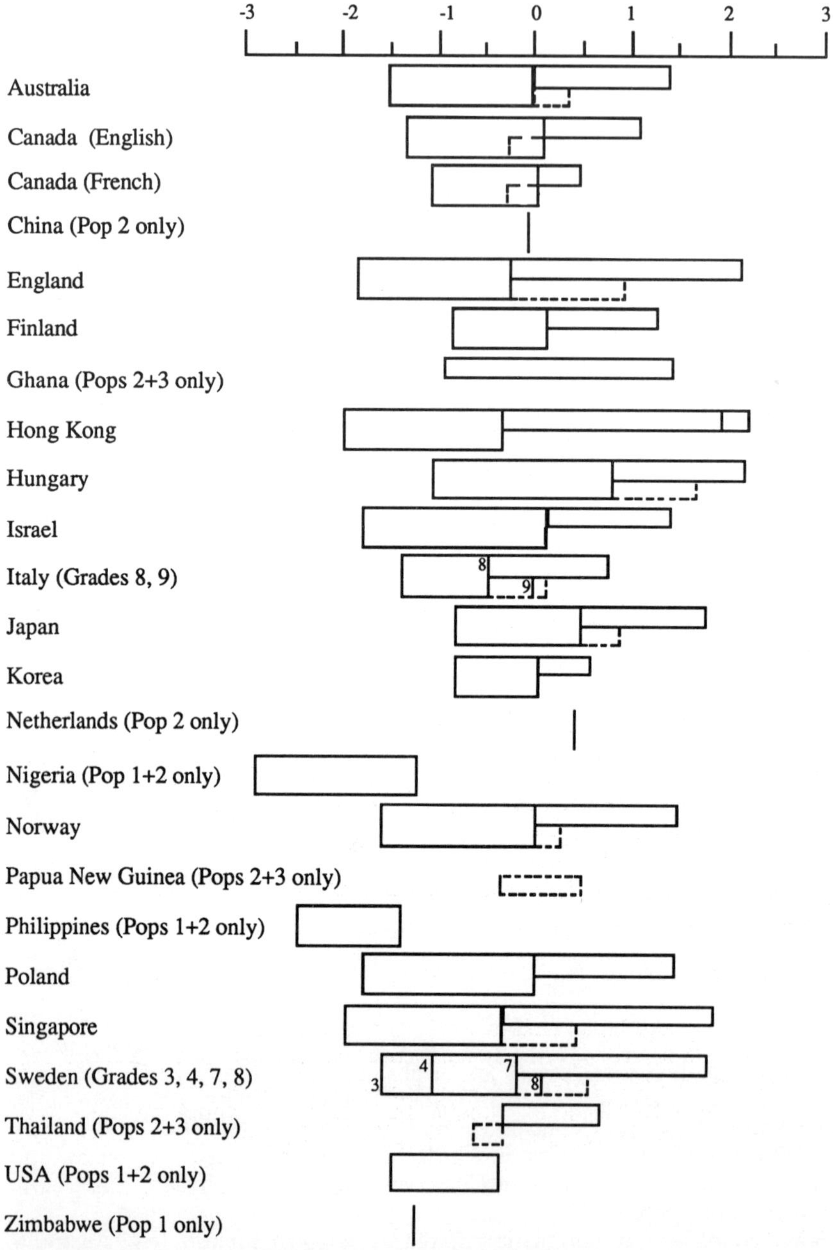

[Figure: horizontal bar chart with x-axis from -3 to 3, showing standard score differences for the following countries/regions:]

- Australia
- Canada (English)
- Canada (French)
- China (Pop 2 only)
- England
- Finland
- Ghana (Pops 2+3 only)
- Hong Kong
- Hungary
- Israel
- Italy (Grades 8, 9)
- Japan
- Korea
- Netherlands (Pop 2 only)
- Nigeria (Pop 1+2 only)
- Norway
- Papua New Guinea (Pops 2+3 only)
- Philippines (Pops 1+2 only)
- Poland
- Singapore
- Sweden (Grades 3, 4, 7, 8)
- Thailand (Pops 2+3 only)
- USA (Pops 1+2 only)
- Zimbabwe (Pop 1 only)

Note: The zero point for scores is the average of all Population 2 scores. The left hand bar is the score of Population 1, the middle bar | for Population 2 and the right hand bars of the students studying science in the terminal grade ⬜ and those not studying science ⬜⎯⎯⎯⎯⎯ . For exact figures see Table 3.11.

It is known that, in general, girls tend not to opt for advanced courses in mathematics and science. However, in this case it is the girls and boys who were in the same classes who were tested. Nevertheless, the participation rates of boys and girls at Population 3 level must be borne in mind.

The above differences are, of course, only gross differences. The issue of why boys perform better than girls in general is complex. Some light is shed on this matter in Chapter 5.

TABLE 3.12 Standard score sex difference [a] in science for all populations

	Population 1	Population 2 (Core + 2 rotated)	Population 3 Biology	Population 3 Chemistry	Population 3 Physics
Australia	.27	.26	.00	.47	.31
Canada (Eng.)	.27	.37	.38	.43	.56
Canada (Fr.)	.27	.47	.07	.16	.23
China	-	.54	-	-	-
England	.23	.37	.25	.31	.06
Finland	.31	.29	.28	.44	.60
Ghana	-	.28	.04	.00	.27
Hong Kong	.22	.34	-	-	-
Hong Kong (Form 6)	-	-	.24	.32	.45
Hong Kong (Form 7)	-	-	.12	.23	.44
Hungary	.13	.16	.20	.68	.39
Israel	.24	.46	-.19	-.52	.51
Italy	.17	-	.56	.11	.39
Italy (Grade 8)	-	.39	-	-	-
Italy (Grade 9)	-	.34	-	-	-
Japan	.12	.31	.25	.37	.26
Korea	.37	.43	.28	.37	.26
Netherlands	-	.52	-	-	-
Nigeria	.18	.29	-	-	-
Norway	.41	.39	.51	.51	.53
Papua New Guinea	-	.45	-	-	-
Philippines	.05	.16	-	-	-
Poland	.26	.36	.25	.55	.68
Singapore	.30	.41	.20	.13	.11
Sweden (3/7/12)	.14	.28	.23	.04	.46
Sweden (4/8)	.30	.41	-	-	-
Thailand	-	.32	.25	.32	.67
U.S.A. [b]	.20	.25	.22	.26	.32
Zimbabwe	-	.29	-	-	-
Average	.23	.36	.24	.33	.40

a   standard score difference = $\dfrac{\text{boys' score - girls' score}}{\text{average sd for all countries}}$
The greater the value the greater is the difference between the sexes.
A value of .00 implies that there is no difference between the sexes.
A negative value means that the difference is favouring the girls.

b   The United States scores are based on core test scores only. All other scores are based on core plus two rotated tests for Populations 1 and 2.

Conclusions

This chapter has reviewed the science achievement test scores of the students at the various levels in the school systems of each of twenty-three countries. This conclusion highlights only the major points emerging from the details given in the chapter.

Population 1

The range of achievement scores was from 35.1 percent in Nigeria, to 66.4 percent in Japan. Korea (with an average class size of 55) scored 65.7. The standard deviation of students was much the same in each country. However, the differences between schools was low in Finland, Japan and Sweden and very high in Nigeria and the Philippines. England, Hong Kong, Israel, Nigeria, the Philippines, Poland and Singapore all had more than 60 percent of their schools with lower scores than the lowest school in Japan. This large variation of achievement is surprising in some countries which pride themselves on 'equality of educational opportunity' and is clearly a point to which those responsible for the national planning of education could give attention.

In terms of subscores in biology, chemistry and physics there is considerable variation within countries. Finland and Korea are the only countries to perform very well across all fields of science. Canada (both English and French) scores well across all subjects. Likewise does Japan with the exception of chemistry. Hungary scores well except for physics. Australia and Norway hold a middle position in all subjects. England is average in biology and earth science but poor in chemistry and physics. Israel, Nigeria, the Philippines and Singapore are poor in all branches of science. It would seem likely that these subject scores reflect the different emphases in the intended curriculum of each country.

However given that there is a very strong relationship between country science achievement at Population 1 and Population 2, it would seem that the ground work laid in the primary grades is very important. The implication is that those countries with poorer science achievement scores at Population 1 could well review their science curriculum at that level.

Population 2

The achievement scores range from 40 percent in the Philippines to 71 percent in Hungary. The Nordic countries and Japan had small differences between schools, but Ghana, Italy (Grade 9), the Netherlands, and the Philippines had large differences between schools. In Ghana and the Philippines this must reflect in part an urban/rural disparity combined with a selective system. Italy has a multiplicity of different school types at the Grade 9 level and the Netherlands, too, has various school types each with a different curriculum. This may reflect a deliberate philosophy as compared with the Nordic educational systems which reflect an egalitarian philosophy.

Hungary is the only country to score consistently well in all branches of science. China (which, it must be remembered, is only the capital and two nearby provinces and, therefore, likely to have better schools than some distant provinces) scores well in all sciences except for biology. Australia and Korea do relatively poorly in chemistry. Thailand performs relatively well in biology, but poorly in the other branches. Italy (Grade 8), England, Ghana, Nigeria, the Philippines and Zimbabwe perform poorly in all branches. It has been noted (Caillods and Postlethwaite, 1989) that the conditions of education in many African countries have been deteriorating and the science scores confirm this. There is a problem. That England performs relatively poorly is a surprise but it must be noted that the English sample is somewhat younger than the samples tested from other countries.

A separate analysis was undertaken on the scores of boys and girls in the upper 20 percent and lower 20 percent of the score distributions in each country. This analysis has not been reported in this chapter because of space restrictions. It was interesting to note that the bottom 20 percent of students in Ghana, Nigeria, the Philippines, and Zimbabwe score at a level just above chance. There is also clearly very low achievement of the bottom group in England, Hong Kong, Italy (Grade 8), Sweden (both grades), Singapore, and the United States. The bottom 20 percent of girls in Sweden would appear to learn no science at Grade 8 when their scores at the end of Grade 7 and Grade 8 are compared. Since the bottom 20 percent of the grade groups are most likely to enter the labor market first, the relatively low scientific literacy of these groups must pose a serious problem for the educational planners in the countries whose scores are very low in this respect.

Population 3

At Population 3 level there are different proportions of the age group in school.

The overall percentage of an age group in school ranges from one percent in Ghana and Papua New Guinea to over 80 percent in Japan, Korea, and the United States. Japan and Korea did not define their target populations to include substantial proportions of the age group enrolled in vocational schools. In biology, the percentage ranges from .2 percent in Ghana to 41 percent in Finland. In chemistry, it ranges from one percent in Ghana and Italy to 37 percent in Korea. In physics, it ranges from one percent in Ghana and the United States to 20 percent in Hong Kong (Form 6). The terminal grade in school is Grade 11 in Canada (Fr.), and Grade 13 in Canada (Eng.), England, Italy, and Hong Kong (Form 7) and Grade 12 in all other systems. The mean ages range from 17 years and two months in Canada and 17 years and three months in Australia to 19 years and two months in Hong Kong (Form 7). The ages of students in most systems, however, are from 18 years old to 18 years and six months. Finally, the average number of subjects studied by students in the various systems ranges from three in England to nine or more in Finland, Hungary, Korea, Poland, and Sweden. In several systems, students took six or seven subjects.

In Hong Kong (Form 7) students study five subjects and their achievement is the highest by a long way in chemistry and physics. Singapore with an average of 5.5 subjects being studied) scores highest in biology (70.9 percent) but Hong Kong (Form 7) follows closely (70.2 percent). Hong Kong (Form 6) scores higher than most countries and in physics scores the same as England (England being Grade 13 and Hong Kong Form 6 being Grade 12). England and Singapore have only about five to six percent of an age group studying the particular fields of a science (i.e. an élite group). However, Hong Kong (Form 7) has 12 percent and Hong Kong (Form 6) had 20 percent of an age group studying science. When the responses to items were classified into the subsets of information, comprehension and application items the profiles reflected the total score. There is a view in many western countries that the Asian countries are very good on factual information but not on comprehension of principles. In the science results presented here, there is no evidence to support this view.

Of note is Japan's relatively low score in biology compared with its performance in chemistry and physics. Canada (Fr.) which is predominantly Grade 11, Italy, Korea and Thailand tend to have the lowest scores across all subject areas. Boys have, on average, higher scores than girls (with the exception of Israel in biology) but the differences are not greater than at Population 2 level for those samples of students taking the science specialist tests which involve different participation rates across countries and subjects.

The between school differences in Hungary, Israel, Italy, Japan, Poland, and Thailand are high. The dramatic shift from low between school differences at Population 1 and 2 to high between school differences at Population 3 in Japan is possibly due to the large number of private schools emerging at this level either for the élite or for those not having passed the entrance examination to Grade 10.

A special analysis for the United States indicates low scores and high differences between schools particularly in biology and chemistry. Thus, neither quality nor equality in achievement is being achieved in science. It should be recalled that the students tested are in a second year science course in senior high school. They are therefore a mixture of advanced placement students and other science students. A study of the élite students in the terminal grade of schooling shows that the élite in Hong Kong (Form 7), and Israel do particularly well in biology and Hong Kong (Forms 6 and 7), and Japan in chemistry and physics.

In general, countries with a higher proportion of an age group in school studying science have lower achievement scores but this does not affect the performance of the élite students.

There is a tendency for those countries where fewer subjects are studied to score higher. The implication, however, is that those systems where only three or four subjects are studied could well consider increasing the number to five or six.

Countries where students are from higher grade levels tend to score higher.

When all of the achievement scores for all populations are put onto one scale the plight of some developing countries becomes very apparent. Some systems seem to take at least one and a half times (in terms of grade years) to reach the same achievement standard as do other countries. In some cases Grade 9 students in some developing countries have a lower mean score than Grade 5 students in some developed countries. In all, there is considerable variation in achievement across countries. Some countries take much more time than others to reach a particular achievement standard.

Boys still tend to score higher than girls. Science would still appear to be considered a 'difficult' subject.

What is possible in science education is demonstrated by Hungary and Japan for mass education (Population 2). What is possible is also demonstrated by Hong Kong at Population 3 level where relatively high proportions of an age group are studying science and where students study five or six subjects on average.

Equality across school in terms of science achievement is demonstrated in the Nordic countries and Japan for mass education at Populations 1 and 2 levels.

Different countries would appear to have different philosophies in terms of the goals of science education and the way in which they organize their school systems. All would appear to be interested in quality and some in equality. The above results raise questions about the relative effectiveness of science education in the countries participating in the study. It is for each country to examine its own results in comparison with those of other countries and review what it needs to do to improve its science education.

References

Caillods, F. and Postlethwaite, T.N. 1989 Teaching and learning conditions in developing countries. In: Caillods, F. (ed.) *The Prospects for Educational Planning*. UNESCO, IIEP.

Doran, R. and Tamir, P. 1992 An International Assessment of Science Practical Skills. *Studies in Educational Evaluation*.

Kish, L. 1965 *Survey Sampling*. John Wiley, New York.

Postlethwaite, T.N. 1991 Achievement in Science Education in 1984 in 23 Countries. In Husén, T. and J. P. Keeves (Eds.) *Issues in Science Education*. Pergamon, Oxford.

4

The Attitude and Descriptive Measures

Any teacher has the problem of not only teaching the cognitive content of science but also of encouraging students to learn science and wanting to learn more. It is also important for students to form attitudes about the non-harmful and beneficial aspects of science for individuals and the society as a whole.

In this study a decision was taken to construct a series of attitudinal and descriptive measures. The attitude scales were concerned with the extent to which the students liked school, had an interest in science, perceived science to have beneficial or non-harmful effects, perceived science as being useful for their future work, the perceived ease or difficulty of learning science, and finally the students' own perception of the amount of effort they made both in classroom learning and in the doing of their homework. The descriptive scales include the extent to which students stated that they participated in experiments and writing up practical work, and the extent to which they perceived themselves as having their own ideas taken into account in the setting and solving of science problems in class. Not all measures were administered at each population level. The following gives a brief overview of the scales (and their directions) administered at each population level. A detailed description of the construction of these ten measures, the means and standard deviations as well as the reliability coefficients are given in Appendix E. It will be noted that some countries did not administer some of the attitude statements within a scale or made an error in the translation of the item. In these cases, means, standard deviations and reliability coefficients were not calculated.

Scales

Practical Work: A high score in Population 1 indicates that the students participated in experiments frequently, either watching the teacher perform an experiment or conducting one themselves. In Populations 2 and 3, a high score indicates that students participated frequently in experiments and did practical work in small groups with written instructions or instructions given by the teacher. Low scores in all populations indicate that minimal practical work occurred in the science classroom.

Beneficial Aspects: In all populations, a high score indicates that students believed science is useful in solving everyday problems, improving living standards and providing for a better future. High scores also indicate that many or most students felt science is worth financial investment and that money funded for science is well-spent.

Facility of Learning in Science: In all populations, lower scores indicate that students perceived science to be a difficult subject to learn, and higher scores indicate more facility in science learning. In Population 1, students were also asked if they had to learn too many facts. A low score indicates those who felt overwhelmed by science data. In Populations 2 and 3, the question of difficulty was expanded to include attitudes to calculations and equipment. Again, a low score indicates those who felt science calculations and equipment to be difficult to employ or use.

Like School: In all three population groups, a higher score indicates more satisfaction and enjoyment of school, including the desire to continue with education.

Interest in Science: Again, a higher score indicates that students were interested in the science taught in schools and found science lessons enjoyable. A low score indicates that students were bored by science lessons.

Student Participation: A higher score indicates that the students perceived themselves as having a say in which topics are studied, that they themselves proposed problems and that the teacher used their ideas. Lower scores indicate the reverse.

The following dimensions were assessed for Population 2 and 3 only.

Career Interest in Science: A high value indicates whether the students considered that science knowledge would be useful in their occupational career and that scientific knowledge would be necessary for their jobs.

Teacher Directed Learning: Students reported about their teachers' approach to instruction. This dimension indicates how the teacher organized the lessons and extent to which the teacher explained the relevance of the topics and helped students when they had difficulties.

The Effort-Scales, **Homework Effort** and **Classroom Effort**, sought to assess the students' motivation at home and in the classroom. Homework effort indicates how completely and thoroughly homework was done. Classroom effort indicates how attentive the student was in the classroom and how much effort he or she took to make up for missed lessons.

Results

One problem in comparing attitude scores across countries is that there is an unknown propensity in some countries for giving 'socially desirable' responses. This can take the form of 'what does my nation expect me to respond' to 'what will please my teacher'. In previous studies it has been contended that this happens in some developing countries where secondary schooling is a privilege enjoyed by a few and where statements on 'liking school' are possibly confounded with the high prestige value, utility, and consequent desirability of schooling. However, it may be argued that in such circumstances the responses to the attitudinal statements reflect accurately a disposition to view all aspects of education highly favourably, since the students are a select group, who aspire to continue with education.

Bearing this in mind, what comments can be made?

The results for Population 1 are not discussed since the scales are very short and thus the reliabilities of some of the scales were low.

The results for Populations 2 and 3 are presented for each scale in the same order as the scales were presented above. In each case, the scale ranges from 1 to 3 except for the last two scales which are from 0 to 3. For the 1 to 3 ranges, *1 = disagree and strongly disagree, 2 = unsure*, and *3 = agree and strongly agree*. For the 0 to 3 scales, the representation is *0 = no effort, 1 = little effort, 2 = some effort,* and *3 = much effort*. In each figure the mean scores for Populations 2 and 3 are

The Attitude and Descriptive Measures

given together except where a country tested only one population and in that case only the results for one population are given.

Practical Work (a descriptive scale)

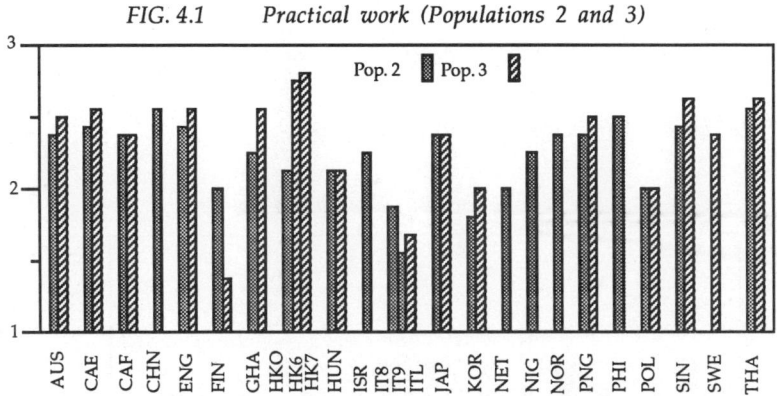

FIG. 4.1   Practical work (Populations 2 and 3)

The extent to which students perceive themselves to be undertaking practical science work oriented activities increases, in general, from Population 2 to Population 3. However, for Canada (Fr.), Hungary, and Japan there was no change. Finland has a dramatic drop from Population 2 to Population 3 level. Italy Grade 8 students report having more practical work than Italy Grade 9 students.

Beneficial Aspects of Science (an attitude scale)

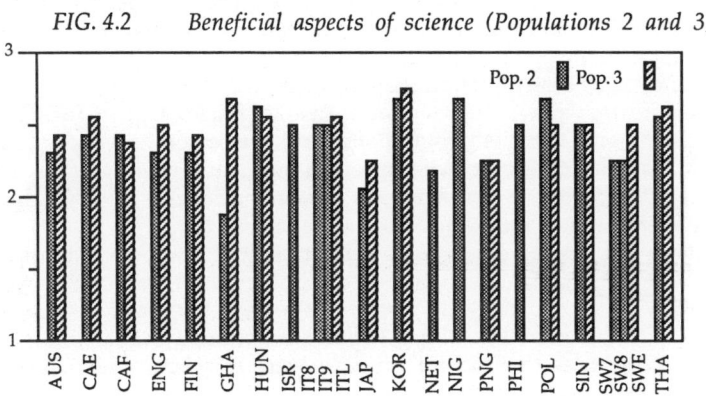

FIG. 4.2   Beneficial aspects of science (Populations 2 and 3)

The Population 2 Ghanaian students regarded science as less beneficial to society than did the Population 3 students. Somewhat more favourable are the judgements in Japan and the Netherlands with scores slightly above uncertain.

All other countries at Population 2 and all countries at the Population 3 level show very positive attitudes towards science. The highest means are found in Korea, Nigeria and Poland at the Population 2 level and at Population 3 in Ghana and Korea. The highest means are in Korea. There is no change between Populations 2 and 3 levels in Papua New Guinea and Singapore. Only in Canada (Fr.), Hungary and Poland are the Population 2 means higher than the Population 3 means. This is likely to be related to the dominance of female students in the schools at the terminal secondary level and in the samples drawn in this investigation for these countries.

Facility of Learning in Science (an attitude scale)

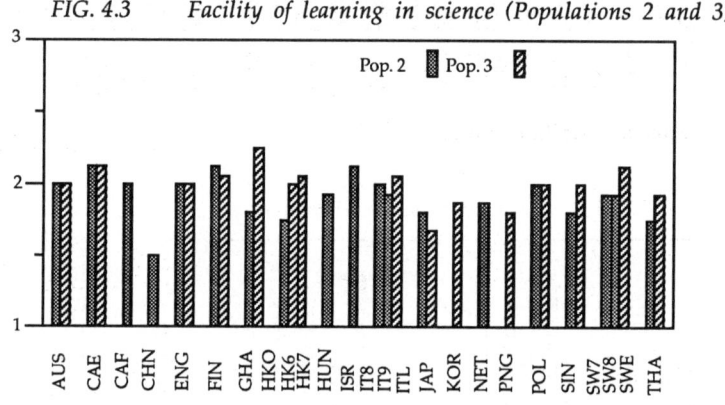

FIG. 4.3   Facility of learning in science (Populations 2 and 3)

A high score indicates that students perceived science as being easy to learn. It can be seen from Figure 4.3 that, in general, students perceived science as being difficult to learn. Only in Finland, Hungary and Japan did Population 3 students perceive science to be more difficult to learn than Population 2 students. Canada (Eng.) and Polish Population 2 and 3 students had the same perception of the ease/difficulty of learning science. In all other countries, Population 3 students perceived science to be easier to learn than Population 2 students.

The countries having a low mean (i.e. disliking school) in Figure 4.4 are Finland, Hungary and the Netherlands and, to some extent, Sweden. Indeed, Swedish eighth graders dislike school more than Swedish seventh graders. It will also be remembered that Swedish eighth graders achieved only marginally higher than seventh graders on the science achievement test.

Population 3 students in the main liked school somewhat more than Population 2 students with the exceptions of Hungary, Italy, Japan, Poland and Singapore. High scores on the scale were obtained for Populations 2 and 3 in Thailand and for Population 2 in China, Papua New Guinea and the Philippines. It was in these countries that not all children were in school and they are all Asian countries where education is prized.

Like School (an attitude scale)

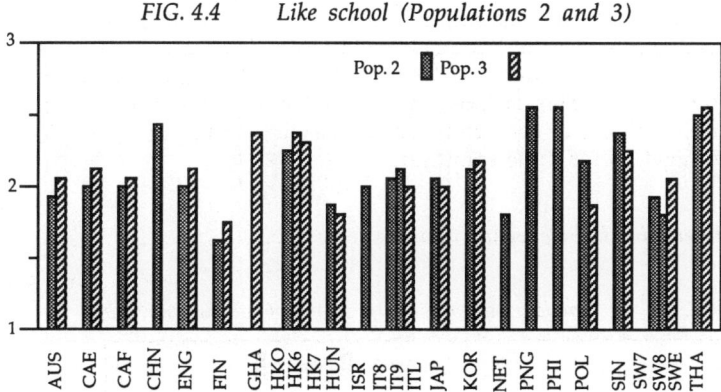

FIG. 4.4 Like school (Populations 2 and 3)

Interest in Science (an attitude scale)

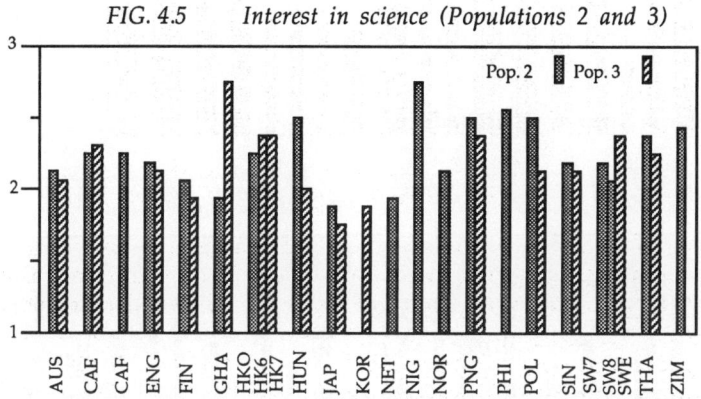

FIG. 4.5 Interest in science (Populations 2 and 3)

A relatively low interest in science as taught in the classroom is reported in Japan and the Netherlands. In eight countries out of 13 Population 2 scores are higher than Population 3 and in five of the countries Population 3 scores are higher than Population 2 scores. Swedish Grade 8 students have less interest in science than Grade 7 students. At Population 2 the between country correlation of interest in science and achievement is -0.46. This is simply because Nigeria, Papua New Guinea, the Philippines and Zimbabwe have higher interest in science scores than countries such as Japan and The Netherlands whereas the inverse was true for achievement. The Ghanaian Population 3 and Nigerian Population 2 values seem to be very high, but it must be remembered that these are a selected group of students.

Student Participation (a descriptive scale)

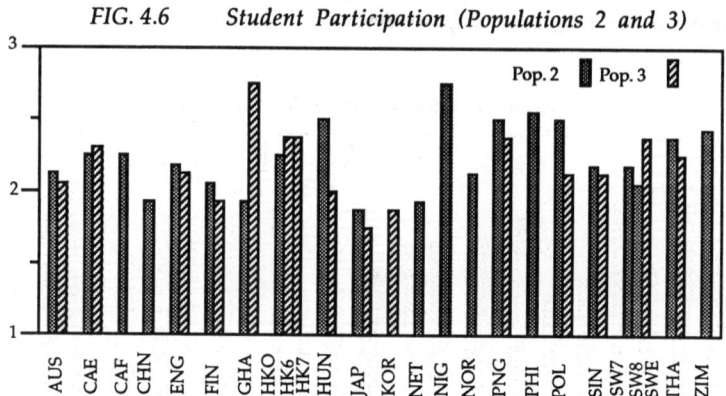

FIG. 4.6 Student Participation (Populations 2 and 3)

Student participation is very low in all countries with the exception of Population 3 in Ghana and Population 2 in Korea. Only in Australia, Canada (Eng.), Ghana and Papua New Guinea do the values increase from Population 2 to Population 3. In most countries there is very little change across populations.

Career Interest in Science (an attitude scale)

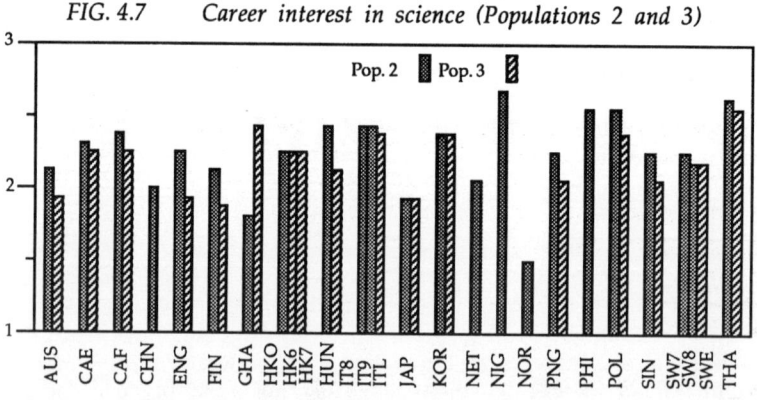

FIG. 4.7 Career interest in science (Populations 2 and 3)

With the exceptions of Hong Kong, Japan and Korea where the same values were obtained at each population level, the perception of a career in science and of the importance of science for a future job decreased from Population 2 to Population 3. Norway (Population 2) has a particularly low value. The relatively low career interest of Population 3 science students in Australia, England, Finland and Japan is surprising and warrants further investigation in those countries. It must be recalled that both science and non-science students are included in the figures and the mean values will depend on the relative numbers in each country.

Teacher Directed Learning (a descriptive scale)

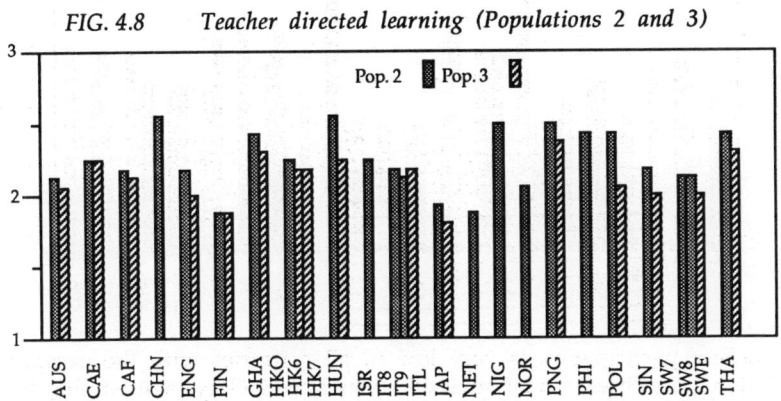

FIG. 4.8   Teacher directed learning (Populations 2 and 3)

Finland, Japan, and the Netherlands have less teacher directed learning than other countries. Population 3 students perceive their learning to be less teacher directed than Population 2 students. These results are in the expected direction in so far as teachers are not so specific in their teaching at the upper secondary school level. The high values for Hungary and the developing countries are, however, an exception to the general rule.

Effort (a descriptive scale)

Only some of the countries administered these scales. Figures 4.9 and 4.10 present the results for 'Homework effort' and 'Classroom effort' respectively.

The homework effort would appear to be low in Finland and Korea (Population 3) and Ghana (Population 2). High values are reported by Hungary, Italy, Nigeria, Philippines, and Poland. Homework effort decreases from

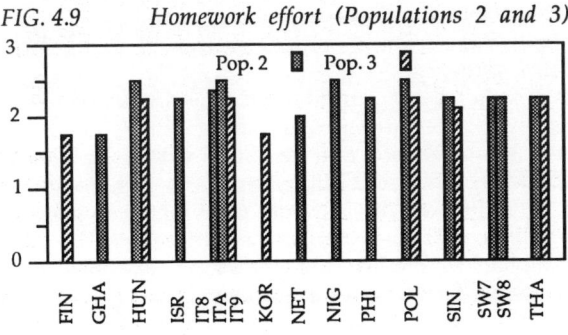

FIG. 4.9   Homework effort (Populations 2 and 3)

FIG. 4.10   Classroom effort (Populations 2 and 3)

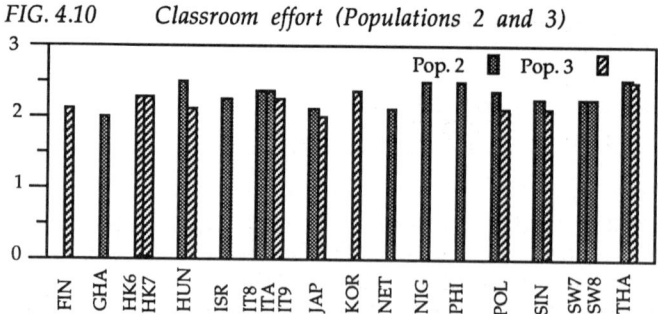

Population 2 to Population 3 level for those countries which administered the instrument to both populations. Classroom effort also decreased from Population 2 to Population 3. The highest values are reported from Hungary, Nigeria and Thailand.

Overall Comment

Not all countries administered all scales at both Populations 2 and 3. This is unfortunate since many of the comparisons are between population levels within countries.

In general most Populations 3 undertook more practical work in science than Populations 2. Given the importance of practical work in science the relatively low values for Finland, Hong Kong, Italy, and Korea (Pop. 2) are surprising and warrant further consideration in those countries.

The fact that students in Population 3 dislike school more than Population 2 students in Canada, England, Hungary, Poland, and Singapore makes one wonder what is occurring in schools to cause this situation. Again, it is surprising and worthy of further study in those countries.

Interest in learning science and interest to enter a job involving science decreased from Population 2 to Population 3 in Australia, England, Hungary, Papua New Guinea, Poland, Singapore, and Thailand. Given the importance of scientific literacy for the increase of the use of technology in the world in general this finding is surprising. Is there something in the career reward system in these countries that is causing this or is it something in the way science is taught? It is perhaps worth noting that in Japan, although students in Population 3 had less interest in learning science, both populations had the same level in terms of career interest. In general, Population 3 students regarded science as more beneficial than Population 2 students.

As will be seen in Chapter 8 the attitudes held by students are a major predictor of science achievement. The attitudes, in most countries, are influenced by classroom effort and liking school and to some extent by conducting experiments in the laboratories. Classroom effort, in turn, is influenced by 'Like School' which, in turn, is influenced by parents. Conducting experiments is influenced by the number of years the teachers have been teaching.

The many interactions among these influences are not easy to disentangle but it is clear that positive attitudes influence achievement.

5

Opportunity and Achievement:
What they tell us about curriculum

In Chapters 2, 3, and 4 information is presented on selected variables concerning students, teachers, and schools and about student cognitive and affective outcomes. In this chapter and the following three chapters the relationships between some of the input, process, and outcome variables are examined. Population 2 was selected for these analyses since the Population 2 students are, in most countries, at a point in the school systems before major drop-out occurs.

What is taught in school represents the opportunity provided for students to learn. For the subjects of earth science, biology, chemistry, and physics and the topics subsumed under each of these science subjects it can be expected that countries and schools will place different emphases on both the various science subjects and their topics.

This chapter examines two measures of opportunity to learn. One of these is derived from pattern characteristics of achievement and is used to assess where the locus of control lies in decisions about the content of what is taught in schools. A second derives from the teacher ratings and is used to assess some of the reasons for country differentiation in achievement levels.

Learning Attitudes, Knowledge and Skill Acquisition and Tested Achievement

The way in which learning activities are organized, both in and out of school, controls students' exposure to and practice on learning tasks. The organization of decisions and actions about learning activities result in the sequencing of learning tasks, the relationships between learning tasks, the amount of time allocated for learning and, in turn, student exposure. In a school context they are known as the curriculum and the teaching-learning process. In out-of-school settings, such experiences come under the rubric of 'socialization' (see, for example, Harnischfeger and Wiley, 1985).

Figure 5.1 summarizes this conception of learning processes and skill acquisition in school contexts and extends it to the testing process. Just as acquired abilities are a result of learning experiences <u>and</u> prior cognitive characteristics of the learners, test-task responses are the outcome of both skills <u>and</u> test-task characteristics (Wiley, 1990).

Thus, three structures can be imposed which influence test item responses.

1. Features of the test tasks.
2. Prior learnings and other characteristics of individuals.
3. The learning experiences of individuals and their organization.

FIG. 5.1 A conception of learning processes, skill acquisition, and test performance

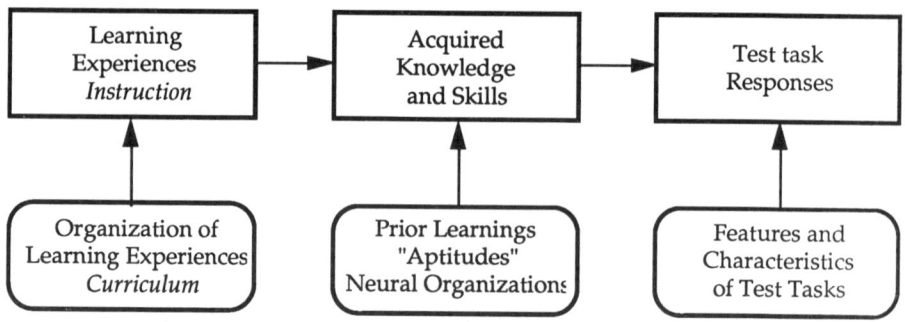

The last two of these have their impact on the skills themselves and their development. Thus, they control the kinds and levels of (and the structure of the relations among) skills acquired. The first, on the other hand, determines the set of skills actually reflected in the tests. Both the skills and their representation in tests influence measured test performance.

Pattern and Levels of Achievement

Knowledge and skills as acquired and as measured have both pattern and level. The organization of the learning activities creates different experiences for students in different countries, schools and classes. Over time, the different learning activities and experiences accumulate and larger ranges of categories of knowledge and skill are learned. These abilities can be accumulated into levels. A student is said to have a high level when he or she possesses many of the knowledges and skills constituting a category. The reverse is true when he or she possesses few. The pattern of learning is the extent to which a student varies in achievement level across categories.

Differences in pattern and level arise because of differences in learning experiences and differences in the abilities and prior learnings which individuals bring to those experiences. The most direct cause of *pattern* differences resides in variations in curricula and in the teaching-learning processes which transform curricula into learning experiences. In the context of this study, these curricular and instructional decisions are said to determine **learning opportunity** in school. These aspects of organization and action are the primary means for controlling the *types* of experiences or opportunities students have for developing the corresponding *types* of skills they acquire. It is across these skill *types* that skill *patterns* are formed. Clearly, these opportunities also influence levels via pacing and instructional efficacy. Pattern and level differences also arise from other sources. Out-of-school learning experiences and differences in the prior learnings and aptitudes of individuals influence skill acquisition. These out-of-school circumstances are likely to have a greater effect on level than on pattern because out-of-school experiences are not specifically planned and coordinated according to the curricular categories used to organize tested achievement.

On the other hand, much of the school curriculum does not focus on skills which are easily acquired outside of school. This is particularly true of mathematics and science. Acquisition of these skills is enhanced or retarded by out-of-school circumstances, such as the family setting, but in most cases these circumstances do not materially contribute to *which* mathematics or science abilities are learned. In other words, out-of-school circumstances contribute primarily to level, not pattern of measured learning.

In-school learning experiences typically occur in groups. Learning experiences which take place outside the classroom, as they relate to the school science curriculum, are usually not group activities. The content focus of out-of-class science learning which is relevant to what has to be learned in school is mostly based on the curriculum through homework assignments. Thus, the experience does not modify the achieved ability *patterns* intended by the curriculum but will generally amplify or reduce their levels because of home-related differences in support for academic learning. Aptitudes, because they enable learning, relate to broad classes of tasks rather than narrow ones. Thus these aptitudes, which are strongly related to home circumstances and background, also have greater effects on the level of skill acquisition than on pattern. The group character of schooling also has important effects. In a given school, the curriculum has many common pattern and level effects on students. Instruction is usually a group activity and, as such, makes for commonality in pattern when implemented on the basis of a planned curriculum.

These structural differences in the effects of in- and out-of-school experiences on individual learning have important implications for knowledge and skills when aggregated across students for a school. The school aggregate will be a mixture of in- and out-of-school effects. Home-related differences in aptitude and home support for school learning, on the one side, and differences in school quality and effectiveness, on the other, produce variations in average *levels* of achievement. Aggregation of these in each of a number of different schools will, in turn, produce school-to-school variations in *level* which reflect community differences in home circumstances as well as school effectiveness.

The consequences for *pattern* aspects are completely different. The group nature of schooling and the planned nature of the curriculum combine to produce similarities in the skill patterns of individuals in the same school or grade program. Out-of-school pattern effects are smaller and much more varied across individuals. Consequently, when aggregated across students in a school, out-of-school effects on pattern are dampened and in-school effects are consolidated. After aggregation, school-to-school variations in achievement *patterns* should primarily reflect differences in the curriculum and in the teaching-learning processes of the schools compared.

The implication of this perspective, for the study of curricular effects on achievement, is a focus on achievement *patterns* rather than levels. Pattern differences should reflect differences in the emphases given by the school to learn the categorized skills. In this case, the skills are science content as reflected in the earth science, biology, chemistry, or physics subtests. As such, these pattern differences form an *indirect* assessment of differences in opportunity. Within countries where schools are permitted to differ in science curriculum one would expect larger pattern variations (e.g., Australia, England, Italy (8 and 9), the Netherlands, the United States). These pattern differences will be returned to later in this chapter. The IEA studies, historically, have used *direct* measures of opportunity to learn. Such direct measures were also used in this study.

Two Measures of Opportunity to Learn

The Direct Measure

The notion of pattern focuses attention on the *content* aspects of learning. Following the lead of previous IEA studies, it is the *opportunity* to learn science which should be one of the crucial mediating factors determining what and how much students learn. The opportunity concept probes the learning experiences of students and attempts to recast them in a way that reveals the knowledge and skills they are supposed to engender. This opportunity concept is extremely important for educational policy and practice. Opportunity to learn is a major social issue because any society has concerns for both equity and adequacy in education. The allocation of opportunity creates both equity and inequity among societal groups. These allocations also preclude or insure adequate education for achievement of societal goals. Opportunities cost resources and require societal investments. Opportunities are necessary prerequisites for learning and economic productivity. That is, learning experiences recast as opportunities to learn particular contents and acquire particular skills have been transformed so that societal assessments can be made of equity, adequacy, and productivity.

The direct operational concept of Opportunity to Learn (OTL) originated in the first IEA studies making international comparisons of achievement (Husén, 1967). Subsequently, all IEA studies have incorporated a *content concept* of opportunity. Most of the emphasis in these studies has focussed on *what* is to be learned (or is 'covered' in instruction) as opposed to how much time is spent on the 'what'. The 'what' in this tradition is not usually defined by a free-standing analysis of all of the 'content' actually covered in the curriculum. Instead it uses the vehicle of a test covering a portion of potential content as a template or 'filter' for mapping curricular coverage by pupils. Also, coverage of curricular content is estimated for an organizational unit rather than an individual student.

In this study an attempt was made, using methodologies developed in earlier IEA studies, to measure the opportunity the teachers in the schools gave their students to learn the items in the tests. Countries varied, however, in their use of the concept. Some participating countries did not collect the Opportunity to Learn (OTL) data at all. Others used one of two distinct questions. These were based on either (a) a question about the percentage of students exposed to the content incorporated in a specific test item, or (b) at what grade the content of that item would first be covered in school. In general, the question was responded to for the school as a whole, not separately for each teacher who taught target grade science students. Often the responses were made by a single 'representative' teacher or group of teachers.

The former type of rating referred to as 'percentage ratings' was considered best employed where students were drawn from more than one class group within a school. The latter type of rating referred to as 'year level ratings' or 'grade ratings' was considered best employed when only one class group of students was selected from within a school. It should be noted that in highly

Opportunity and Achievement

centralized school systems there is likely to be very little variation between school and classroom groups in the ratings assigned by the teachers of the students tested because of the demands to conform to the centrally prescribed curriculum.

Percentage Ratings. For the percentage ratings the teachers (or group of teachers) were asked, with reference to the group of students in the sample from their school, to indicate for each item on each test:

a) All or most (at least 75 percent) of the group of students have had the opportunity to learn the concepts tested by the item.
b) Some (25 percent to 75 percent) of the group of students have had the opportunity to learn the concepts tested by the item.
c) Few (fewer than 25 percent) of the group of students have had the opportunity to learn the concepts tested by the item.
d) None (0 percent) of the group of students have had the opportunity to learn the concepts tested by the item.

This question was used where sampling occurred across classes.

Grade Ratings. For the year level ratings the class teacher was asked, with reference to the group of students in the sample from their school, to indicate for each item on each test the year level (grade level) at which the concepts tested by the item was first taught to the group of students tested.

Response Code	Population 1	Population 2	Population 3
A	Year 4 or earlier	Year 8 or earlier	Year 10 or earlier
B	Year 5	Year 9	Year 11
C	Year 6 or later	Year 10 or later	Year 12
D	Not at all	Not at all	Year 13
E			Not at all

This question was used where intact classes were tested.

In some countries, there was no OTL data or it could not be deciphered. A simple recoding of the OTL information was made in 15 countries. In each of these, the OTL response was dichotomized in an effort to determine whether typical students did or did not have the instruction needed to answer the question. The percent question was interpreted as indicating OTL existed when 75 percent or more of the students was said to have had appropriate instruction and the grade question was interpreted as indicating OTL existed when the content was said to be from the current grade or an earlier grade. The grade question was used in China and Hungary and the percent question in the other countries.

The Indirect Measure

Earlier, mention was made of pattern and level effects on achievement. A focus on pattern effects as an indirect assessment of learning opportunity was advocated. The procedures for the direct measurement of opportunity were mentioned above. As will be seen later in this chapter, the direct measurements must be used with caution as they could be distorted by measurement errors deriving from ambiguous understandings and lack of knowledge on the part of the respondents. An alternative is the indirect measurement by means of assessments of achievement *patterns*.

Level of achievement can be separated from *pattern* of achievement. In Table 5.1 below, some hypothetical achievement values are presented. The column designations (item or topic) denote some sub-categories of knowledge or skills measured in the testing process. These could be particular test items or groups of items such as subtests. The entries in the table are then achievement 'scores' of some kind. These are typically item difficulties or average numbers of items correctly responded to in the skill category. The row designation (school) corresponds to the entities to be compared with respect to curriculum or opportunity. Each row is an achievement *profile* for a particular school.

The entries at the foot of each table (the row labelled 'mean') are the means of the values for each item or topic. The entries to the right side (the column labelled 'mean') are the means of the values for each 'school'. These latter means are taken as a measure of 'level' as they reflect average achievement over the items or topics. In order to remove the level effects, they are subtracted from each of the values in their respective rows. The '20' in row 1/column 1 of Panel 2, for example, equals 75 minus 55, which is the corresponding value and the row mean from Panel 1. Thus, the rows in Panel 2 are *level-adjusted* profiles and represent achievement *patterns*. It can be seen in the example, that School 1 and School 2 have identical level-adjusted profiles and, therefore, the same estimated patterns. However, these schools have a considerable *level* discrepancy (55 vs. 35). The inference from this is that these schools have similar curricula and opportunity structures, but considerable difference in effectiveness or clientele, or both.

TABLE 5.1   *Illustration of profile adjustment and dissimilarity calculation*

Panel 1				
Profile				
Item or topic				
School	1	2	3	Mean
1	75.0	35.0	55.0	55.0
2	55.0	15.0	35.0	35.0
3	35.0	75.0	55.0	55.0
Mean	55.0	41.7	48.3	48.3

Panel 2				
Level adjusted Profile				
Item or topic				
School	1	2	3	Mean
1	20.0	-20.0	0.0	0.0
2	20.0	-20.0	0.0	0.0
3	-20.0	20.0	0.0	0.0
Mean	6.7	-6.7	0.0	0.0

Panel 3				
Item deviations				
Item or topic				
School	1	2	3	Mean
1	13.3	-13.3	0.0	0.0
2	13.3	-13.3	0.0	0.0
3	-26.7	26.7	0.0	0.0
Mean	0.0	0.0	0.0	0.0

Panel 4				
Dissimilarity calculation (mean squares)				
Item or topic				
School	1	2	3	Mean
1	177.8	177.8	0.0	118.5
2	177.8	177.8	0.0	118.5
3	711.1	711.1	0.0	474.1
Mean	355.6	355.6	0.0	237.0

Root Mean Square Dissimilarity Index: 15.4

Opportunity and Achievement 97

Panels 3 and 4 carry this issue in the direction of <u>dissimilarity</u> assessment. One important topic to be explored below is the degree to which schools (or countries) have similar curricula. If all of the schools in a country have the same curriculum and if that curriculum is implemented in a consistent fashion across all of the schools in the country, then the learning opportunities should be similar and consequently the resulting skill or achievement patterns should also be similar. In the context of Panel 2, the level-adjusted profiles should all be the same. In fact, since these adjusted profiles would be the same, they would all be equal to their column means. Therefore, if the column (item or topic) means were subtracted from the adjusted values the result would be zero for all cells in the table. In the example given, this is only true for Item 3. The reason for this is that the adjusted profile for School 3 was not the same as that for Schools 1 and 2. In Panel 4 indices of dissimilarity have been calculated for this set of schools by squaring the values in Panel 3. These squares are averaged in the row and column labeled 'mean'. The resulting overall mean square value is 237. This value would be zero if all of the adjusted profiles were the same. The more discrepant the adjusted profiles, the larger the mean square will be. Another index can be calculated by taking the square root of the mean square. The value of this so-called *Root Mean Square Dissimilarity Index* is given below the panels. Both kinds of overall indices will be used in the analyses reported below.

Curricular Control and Curricular Similarity

The variation in opportunity to learn is, in part, a function of the decision making about what is taught. Decision making about the content of science education was assessed in this study by a question on the loci at which decisions are made. Table 5.2 presents school principals' responses about these loci. Thus, in Australia 22 percent of the principals said that the decisions were made at a more central level than the school. (In Australia this probably means at the state level). One percent said the decisions were made at the school level, one percent 'other' which probably means department head, and 76 percent by the teachers in the school. Thus in Australia decisions are split between central authorities and individual teachers.

It can be seen that in five countries, Finland, Hungary, Japan, Papua New Guinea, and Poland, content decisions are centralized at a regional or national level. In one country, Italy, such decisions are made by teachers. Also, in England almost every school indicated teacher authority on science curriculum matters. In all other countries, patterns of decision making vary from school to school, some decisions are centralized, some made at the school level, and some by teachers. The only unusual pattern is in the United States where 'other' individuals or groups make content decisions in 43 percent of the schools. Since American high schools often have science departments with chairpersons, this may mean that many such decisions are made at this intermediate level in these schools. A more likely alternative is that the 'district' structure of American school governance implies many decisions are made at levels below the region, but above the school.

These differences in locus of decision making should have important consequences for opportunity to learn. In particular, countries in which the curriculum is nationally decided should have similar opportunity and achievement patterns. In other words, the level-adjusted profiles should be close to being equal. In this case the mean square index should be small. This might also be the case if there is a strong consensus among those making the decisions, even when the decisions are dispersed. If, on the other hand, decision making is dispersed and those making the decisions do not hold common values or views

TABLE 5.2 Levels of decision about science content, by country

Country	Regional or Central Authority	School	Other	Teacher
Australia	22	1	1	76
Canada (Eng.)	74	10	1	15
Canada (Fr.)	84	9	5	2
England	0	1	0	99
Finland	100	0	0	0
Ghana	49	6	0	44
Hong Kong	13	15	1	72
Hungary	100	0	0	0
Israel	49	12	0	38
Italy (Grade 8)	0	0	0	100
Italy (Grade 9)	0	0	0	100
Japan	100	0	0	0
Korea	59	7	0	34
Netherlands	15	7	0	78
Nigeria	43	20	0	38
Papua New Guinea	100	0	0	0
Philippines	68	17	1	14
Poland	100	0	0	0
Singapore	53	6	0	41
Thailand	86	3	0	11
U.S.A. (Phase 1)	7	10	43	40

about what should be learned, then opportunities should vary and the index should be large. Also, even if curricular decisions are nominally centralized, when there exist no control mechanisms for checking and enforcing conformity, diversity may also result.

The mean square dissimilarity index[1] for countries was calculated. This consisted of calculating school level item difficulties for each four science subject subsets of items: biology, chemistry, earth science, and physics. Each subset of difficulty values served as a profile. Each set was analyzed for each country, resulting in a within subject school mean square dissimilarity index for each of the four science subjects in each country. These results were checked in a separate analysis not reported here. In this analysis the difficulties were summed within each subset and the four subject sums were used as a profile for each school. This analysis confirmed the results shown below.

The item mean square dissimilarity indices are displayed in Table 5.3.

Within the science subjects, and for almost all countries, school similarity was greatest for physics and least for earth science. This probably reflects a settled international as well as country specific consensus about physics compared with earth science content.

It can be seen that the greatest similarity amongst schools occurred in Finland, Japan, Norway, and Sweden and the least similarity amongst schools in

---

[1] Because of the incompleteness of the data design, a special procedure was constructed for calculating these indices. This is outlined in Wolfe (1989).

Ghana, the Netherlands, Nigeria, and Singapore. An analysis[1] of the values in Table 5.3 indicated that most of the variation amongst these indices was between countries and between science subjects. There was very little deviation from an index value jointly projected on the basis of an average country and average science subject indices.

TABLE 5.3 Item-school dissimilarity indices, by science subject and country

Country	Biology	Chemistry	Earth Science	Physics	Mean
Australia	6.1	9.4	9.3	4.1	7.2
Canada (Eng.)	6.8	14.6	14.4	5.4	10.3
Canada (Fr.)	6.9	10.1	11.9	4.3	8.3
China	8.5	14.7	15.8	7.2	11.6
England	6.3	12.0	9.0	4.0	7.8
Finland	2.1	3.8	5.4	1.9	3.3
Ghana	19.1	30.0	41.4	19.0	27.4
Hong Kong	9.1	12.4	20.7	9.4	12.9
Hungary	11.6	17.4	17.5	10.1	14.2
Israel	14.1	31.9	25.5	13.7	21.3
Japan	2.6	3.3	3.4	1.1	2.6
Korea	5.6	9.1	14.0	5.4	8.5
Netherlands	13.5	40.5	30.9	15.5	25.1
Nigeria	23.6	34.1	46.4	22.4	31.6
Norway	5.0	5.4	12.9	2.2	6.4
Papua New Guinea	5.7	6.3	11.4	3.4	6.7
Philippines	18.3	22.8	40.6	15.4	24.3
Poland	13.8	27.7	33.0	15.8	22.5
Singapore	19.9	38.4	28.6	13.2	25.0
Sweden (Grade 7)	3.0	6.9	7.5	3.6	5.2
Sweden (Grade 8)	5.4	7.5	8.9	3.8	6.4
Thailand	5.2	10.1	16.2	5.7	9.3
U.S.A. (Phase 1)	10.9	24.3	20.5	8.2	16.0
Zimbabwe	6.8	12.5	14.9	8.9	10.8

In Table 5.2, it was seen that the locus of control over what content was to be taught lay above the control of the school for both Finland and Japan. (Unfortunately, neither Norway nor Sweden provided data). Ghana had more or less an equal split between teachers and the regional or central authority; the Netherlands had 78 percent of the schools where the teacher had authority and only 15 percent where the decisions was made above the school level. In both Nigeria and Singapore, there was more or less an equal split between schools reporting that teachers had the most say or the regional or central authority had the most say. It may be that in these countries, part of the diversity in learning opportunity comes about because the decision processes above the school result in different priorities than those taking place within schools. If the values and

---

[1] Logarithmic transformations were made of the dissimilarity indices and the transformed values were subjected to an analysis of variance. In the resulting analysis, both the country and science subject effects are large in comparison to the interaction mean square.

views of the teachers differ substantially from those of regional or central authorities, this could be a powerful source of diversity in opportunity.

Two anomalies in this comparison are Poland and Korea. Poland reports that curricular decisions for all schools are made at levels above the school. However, school dissimilarity in that country is particularly great, exceeding that in Israel and the United States. One possibility is that although curricula are nominally set above the school level, control processes are weak. Korea, on the other hand, reports a decision-making split similar to Singapore, and tends towards a similar pattern across schools in item difficulty. It may be that there are social mechanisms of curricular consensus in Korea that cut across institutional and administrative lines.

It is also noteworthy that countries with relatively high degrees of decentralization of decision making to teacher, such as England (in 1984 when the data were collected) and Hong Kong, do not necessarily have high degrees of school dissimilarity. It may be that either teacher or central decision making promote some degree of pattern similarity, the one through consensus and the other through control. Only relatively sharp splits, with substantial numbers of schools reporting decisions made above the school level combined with many other schools reporting teacher decisions, may indicate both a lack of consensus and a lack of central control. One key may be whether the decisions made above the level of the school are made by national, regional, or local jurisdictions, covering a limited number of schools. Unfortunately, the questionnaire items are not specific enough to allow such comparisons.

In addition to the above analyses of school by item dissimilarity, an estimate was also constructed of the school by science subject dissimilarity index for each country. The general results are parallel to the school by item analysis[1]. Japan and Finland have the greatest similarity over schools and Nigeria and Ghana the least. In the main, the pattern of curricular homogeneity at the item level is very similar to that at the science subject level. This is consistent with the notion that the processes of consensus or control of subject emphases and of subtopic emphases are similar.

Opportunity and Achievement

The Distribution of Direct Measures of Opportunity to Learn

The OTL ratings were averaged over teachers and over items within each of the four main subjects, and the results are shown in Table 5.4.

Opportunity to Learn is actually a difficult concept for many teachers to deal with (what is the knowledge or skill needed to answer a particular item?) since understanding of skill and knowledge domains is incomplete. Another issue is that the knowledge of what has been taught may not be fully possessed by those responding to the question. For example, the previous grade in some countries may be in another school and the teachers in the school with the target grade may not know whether or not some content has been taught. In addition, the exact formulation of the OTL question may interact with that concept and elicit responses with differential meaning. However, the fact that they are lower for

---

[1] The correlation between the country effects from the logarithmic analysis of school by item components and the log school by topic components is 0.77.

Italy Grade 9 than Italy Grade 8, while Grade 9 is the population of older students, indicates that some problems exist.

TABLE 5.4 *Average percent Opportunity to Learn by subject and country*[1]

Country	Biology	Chemistry	Earth Science	Physics
Australia	45	38	51	50
Canada (Eng.)	41	61	38	54
Canada (Fr.)	44	19	34	23
China	85	78	70	80
England	43	43	38	51
Hungary	94	90	94	96
Italy (Grade 8)	47	23	46	42
Italy (Grade 9)	34	17	35	31
Japan	64	61	62	69
Korea	35	24	38	36
Netherlands	28	29	38	43
Nigeria	31	26	26	26
Philippines	56	48	43	44
Singapore	38	40	29	45
Thailand	64	53	62	63

The reader should recall that OTL, in the sense defined here, means opportunity to learn the content incorporated in the science tests used in this study. It does *not* necessarily reflect the priority content in a country's core curriculum. In constructing the tests, a concerted effort was made to incorporate the content common to most curricula. However, it was not possible to represent each country's priorities equally well in the tests. In what follows, OTL and achievement subscores, are analyzed by science subject, to make for more meaningful conclusions. Results are reported by country and science subject. In addition, achievement scores are related to measured OTL. These relations are revealed by a series of analyses given below.

The data in Table 5.4 allow a two-way breakdown of opportunity means. They were analyzed by decomposing the values into four parts: (a) an overall average value, (b) a science subject effect - the difference between the science subject average and the overall average, (c) a country average - defined similarly to the science subject effect, and (d) the discrepancy between the original values given in Table 5.4 and the value expected on the basis of (a), (b), and (c). This analysis was performed by means of a technique know as 'median polishing' (Tukey, 1977)[2]. This kind of analysis is less subject to the distorting effect of outlying or errorful values than typical analyses of means. Below, use will be made of more standard, mean-based analyses to assess the precision of the results.

Displayed in the three panels of Figure 5.2 are, in turn, the distributions of effects of (a) the science subjects, (b) the country, and (c) their discrepancies (interactions). The 'dots' in these panels represent the effect values. The scale is

---

[1] Finland was inadvertently omitted from this analysis.
[2] This procedure involves the iterative calculation of the medians of the differences between row values in the same column and vice versa.

the same for all panels. Figure 5.2 shows that in addition to the large country effects (middle plot), there are rather small overall subject effects (top plot), and only three rather large country by subject interactions (bottom plot). This latter result implies that these countries differ in the relative emphasis they give to the four science subjects.

FIG. 5.2   Distribution of science subject, country, and interaction effects in Opportunity to Learn ratings

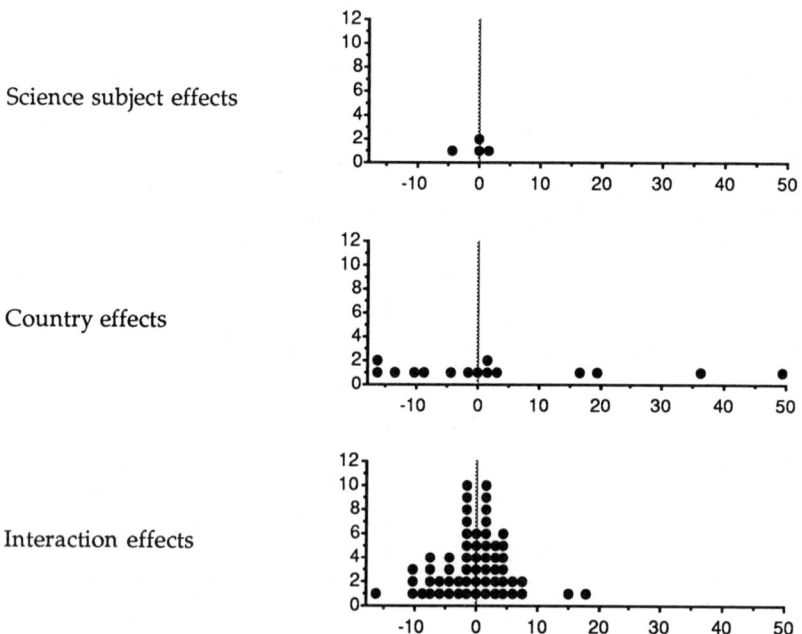

Science subject effects

Country effects

Interaction effects

For example, Canada (Eng.) has, in comparison to its emphasis on other subjects, a relatively higher opportunity to learn chemistry, while Canada (Fr.) has, again in comparison to its other subjects, higher opportunity to learn biology. Earth science shows relatively low OTL in Nigeria and Singapore. Chemistry opportunity is relatively low in both the Italian populations.

These comparisons of opportunity across science subjects assume that opportunity judgements for the four topics had the same basis *within* each country. The meaning of *between* country comparisons of OTL requires more careful analysis. In the next section, an analysis of the relation between the directly measured OTL and achievement across countries and science subjects is undertaken. In addition to the intrinsic interest in finding out how much differential achievement between countries is due to differential opportunity, this analysis also forms an important cross-validation of the OTL percentages themselves.

Relations of Learning Opportunity to Achievement

The simple notion behind relating opportunity to achievement is that students will perform better on topics for which they have received instruction. Average performance on a topic will be relatively higher in countries which give relatively higher opportunity to learn that topic.

One feature of the science achievement tests is that the items are not matched in level of difficulty across topics. For the 15 countries with OTL data, the mean overall percentages of correct achievement are 56.8 (biology), 49.1 (chemistry), 63.2 (earth science), and 59.5 (physics). These differences in average score reflect (a) variations in complexity of the abilities required together with differential effectiveness in the teaching and learning of the science subjects; and, (b) variations in opportunities to learn the content and skills. The several potential sources of these differences make it difficult to compare the relations between opportunity to learn and achievement across science subjects.

Country differences in achievement may also be thought of as deriving from two sources: (a) country differences in opportunity to learn the content covered by the test, and (b) country differences in process of teaching and learning the covered content. By taking opportunity to learn into account, it is possible, at least partially, to disentangle which parts of country or science subject differences in achievement are due to opportunity and distinguish them from those directly connected to teaching and learning.

As a first approach on the opportunity-achievement relation, an analysis of covariance was performed in which the 60 (15 countries by four science subjects) achievement values were related to the 60 OTL values. The analysis resulted in a regression relation of achievement to opportunity and two sets of adjusted achievement means, one for subjects and the other for countries. These values reflect subject and country differences after adjusting for opportunity. The slope (raw regression coefficient) relating achievement percentage to OTL percentage was .136 indicating that a 10 percent difference in OTL corresponded to a 1.36 percent difference in achievement.

The summary table of the analysis of covariance is displayed in Table 5.5.

TABLE 5.5 Analysis of covariance

Source	df	Sum of Sq.	Mean Sq.	F-ratio	Prob.
OTL	1	34	34.1	4.23	0.046
Subject	3	1356	451.8	56.07	0.000
Country	14	2886	206.1	25.58	0.000
Error	41	330	8.1		
Total	59	5864			

The table exhibits a statistically significant regression relation between OTL and achievement. The size of the slope, discussed above, is relatively small. The slope estimate of .136 is surely smaller than the true value since the OTL values clearly have considerable measurement error. The differences between topics and countries are quite large after adjustment for OTL, indicating considerable variation in teaching-learning processes, net of opportunity.

The original and adjusted achievement means, by science subject, are presented below in Table 5.6.

TABLE 5.6 Raw achievement means by science subject and adjusted means for Opportunity to Learn

Subject	Raw Mean	Adj. Mean	Adjustment
Biology	56.81	53.46	-3.35
Chemistry	49.08	46.56	-2.52
Earth Science	63.23	59.73	-3.50
Physics	59.50	68.87	9.37
Mean	57.15	57.15	0.00

The differences between the adjusted and the raw means are negative for biology, chemistry, and earth science and positive for physics. This corresponds to the fact that the opportunity to learn the tested physics content was less than the opportunities for the other subjects. After adjustment, the achievement differences across the science subjects are even greater than before. Physics replaces earth science as the topic with the highest achievement percentage. This could either mean that the physics test items are less complex than those in the other areas or that instruction in physics is actually better than that on the other subjects.

Table 5.7 displays the corresponding values for countries for all science subjects together.

The adjustments indicated in the table are rather large for two countries: Hungary and China. After accounting for the fact that these two countries have uniformly higher opportunity ratings than the others, their average adjusted achievement means are lowered. Hungary now shares its high achievement - net of differences in opportunity - with Italy (Grade 9) and Japan. China is now in a second ranked group along with Canada (Fr.) and Singapore. Thus, an important part of the high achievement standing of Hungary and China is due to the fact that their curricula - as delivered by teachers - better matches the test content than the curricula of the other countries. To the extent that the tests validly represent the important core curricular content this is favorable to the science education in Hungary and China: to the extent that the test represents content not in the desired core, the higher raw achievement values for these countries are spurious.

By using the estimates from the analysis of covariance, 60 values can be projected for achievement from opportunity in the 15 by 4, country by science subject array. These values are displayed in the left side of Table 5.8, along with the discrepancies (actual minus projected) between them and actual achievement levels (right side). These values represent cell by cell variations, rather than row by row or column by column variations in opportunity.

Opportunity and Achievement

TABLE 5.7 *Total raw achievement means by country and adjusted means for Opportunity to Learn*

Country	Raw Mean	Adj. Mean	Adjustment
Australia	60.25	60.47	.22
Canada (Eng.)	60.50	60.38	-.12
Canada (Fr.)	57.25	59.65	2.40
China	63.25	59.07	-4.18
England	55.50	56.02	.52
Hungary	70.50	64.24	-6.26
Italy (Grade 8)	52.50	53.60	1.10
Italy (Grade 9)	62.00	64.50	2.50
Japan	66.50	64.26	-2.24
Korea	61.50	63.46	1.96
Netherlands	60.25	62.04	1.79
Nigeria	40.25	43.02	2.77
Philippines	40.25	40.23	-.02
Singapore	57.75	59.06	1.31
Thailand	56.50	54.74	-1.76
Mean	57.65	57.65	0.00

TABLE 5.8 *Projected achievement means based on Opportunity To Learn, and discrepancies between actual and projected (by country and science subject)*

Country	Projected				Discrepancies			
	Bio	Chem	Earth	Phys	Bio	Chem	Earth	Phys
Australia	59.3	51.5	66.4	63.7	+0.7	-1.5	+1.6	-0.7
Canada (Eng.)	58.7	54.5	64.6	64.2	+4.3	-3.5	+1.4	-2.2
Canada (Fr.)	58.4	48.1	63.3	59.2	+0.6	-1.1	-1.3	+1.8
China	63.4	55.5	67.6	66.4	-9.4	+7.5	-0.6	+2.6
England	54.6	47.7	60.2	59.4	-0.6	-0.7	-0.2	+1.6
Hungary	69.8	62.4	76.1	73.8	-1.8	+1.6	+1.9	-1.8
Italy (Grade 8)	52.8	42.6	58.9	55.8	+0.2	-0.6	+1.1	-0.8
Italy (Grade 9)	61.9	52.7	68.3	65.2	+0.1	+0.3	+0.7	-1.2
Japan	65.7	58.4	71.7	70.1	-0.7	-1.4	-1.7	+3.9
Korea	61.0	52.6	67.7	64.8	-1.0	-0.6	+0.3	+1.2
Netherlands	58.6	51.8	66.2	64.3	-2.6	+1.2	-0.2	+1.7
Nigeria	40.0	32.4	45.6	43.0	+5.0	+0.6	-2.6	-3.0
Philippines	40.6	32.6	45.1	42.7	+1.4	+1.4	-0.1	-2.7
Singapore	57.0	50.4	62.0	61.6	+0.0	-1.4	+1.0	+0.4
Thailand	56.2	47.8	62.2	59.8	+3.8	-1.8	-1.2	-0.8

Scrutinizing these values, it can be seen that projected values vary considerably. In fact, the three factors of country, science subject, and OTL account for all but one or two tenths of one percent of the achievement variation over countries and subjects. Still it can be seen that there are remaining discrepancies of significant magnitude. These are large for China, for which biology achievement is more than nine points less than projected and chemistry is over seven points greater. Canada (Fr.) has the opposite pattern to a somewhat smaller degree. Japan has higher than projected achievement in physics, and Nigeria and Thailand in biology.

These results give rise to the question of whether the relation of OTL to achievement varies by subject or country. Some of these large discrepancies might be accounted for if the slope of the relation really varied by either topic or country. This was investigated by augmenting the analysis of covariance in two distinct ways: (a) allowing slopes to vary by science subject, and (b) allowing them to vary by country. The outcomes of these analyses gave no support for slope variation by topic (an F-ratio of .17 with 3 and 38 degrees of freedom). The outcomes for variation by country were somewhat stronger (an F-ratio of 1.53 with 14 and 27 degrees of freedom).

Conclusions

Only Population 2 data were examined in this chapter. An examination was undertaken of the similarity of achievement profiles amongst schools. This was done using item difficulties for each of four science content areas and with summary measures across the four areas. Profile similarity indicates whether the curricula implemented in a country's schools provided closely aligned learning opportunities. The profiles were most parallel across schools in the Nordic countries and Japan and least in Ghana, the Netherlands, Nigeria, and Singapore. The large differences in the kinds of learning opportunities among schools in these four countries is not surprising since Ghana had two distinct school types in the sample with different curricula, as the Netherlands had several school types with different types of classes with different curricula and Singapore sampled different types of classes within schools and these types have different curricula.

Content decisions are seldom made at the school level in any country. That is, most of the decisions are either made by teachers or at system levels above the school. Where the decisions about the content to be taught are taken at the regional or national level of the school administrative system then there is more similarity in achievement profiles between schools. Where decisions on content are made by individual teachers within schools (i.e. in decentralized systems) then there is somewhat less similarity in these profiles between schools. The greatest profile dissimilarity in achievement between schools is associated with the split decision responsibility - some above the school level and some by teachers.

An attempt was made to measure the opportunity each class had had to learn the content and concepts embodied in the items in the test. Chinese and Hungarian students had the highest opportunity and Nigeria the lowest. Most countries had similar amounts of opportunity across the four subject areas. However, Canada (Fr.) had lower opportunity in chemistry and physics than in biology and earth science, and Italy had lower opportunity in chemistry than in the other science subjects.

The measured OTL had a low to moderate relationship with achievement within countries. It can be argued that if the measures of OTL had been more accurate then a stronger relationship would have been found. At the between country level, there was a stronger relationship between OTL and achievement.

If adjustments are made to the raw country scores by taking account of the opportunity to learn the test content then there are some changes to the scores but with the exception of China [see Table 5.7], this does not significantly affect the rankings of the countries.

What does all of this mean for educational planners? The implications may be that if planners wish to have less variation between schools in science achievement then it is wise to define the content to be studied nationally or regionally. What is not clear from these analyses is the implication of such a policy for achievement *levels* as opposed to *patterns*. Where centralization is not possible or desirable then school dissimilarities in emphasis coming about from allowing teachers to select the content may be tempered by teacher training and textbook choice.

Opportunity to learn is important across countries and planners in low scoring countries should re-examine which science and the amount of science they are asking their students to learn.

References

Harnischfeger, A. and Wiley, D. E. 1985 *Origins of Active Learning Time*. In: Fisher, C. and Berliner, D. (Eds.) *Perspectives on Instructional Time*. Longman, New York.

Husén, T. (Ed.) 1967 *International Study of Achievement in Mathematics*. (Vols. I & II) Almqvist and Wiksell, Stockholm.

Tukey, J.W. 1977 *Exploratory Data Analysis*. Addison-Wesley, Reading, Mass..

Wiley, D. E. 1990 Test validity and invalidity reconsidered. In: Snow, R. and Wiley, D. E. (Eds.) *Improving inquiry in social science*. Erlbaum, Hillsdale, New Jersey. Also in earlier version (1987) as *Studies of Educative Processes*, No. 20, Northwestern University, Illinois.

Wolfe, R. G. 1989 *A hierarchical variance-covariance component decomposition of test performance: Schools, pupils, and items*. Computer documentation, Hamburg.

6

Teachers: Teaching Loads, Grades Taught, and Subjects Taught in Population 2

In Tables 2.7 and 2.8 of Chapter 2 information was given about the teachers and about the schools. This chapter examines the teachers of science at Population 2 in more detail. It looks at the teaching load, the grades taught, the subjects taught, and the relationship between the subjects and grades taught. The emphasis is on science instruction as delivered.

Table 2.6 in Chapter 2 presented summary information on student characteristics. One item of information concerns science instruction. The number of hours of science instruction per week for a student ranged from three hours in Thailand to 7 hours in Poland. The students in some countries all took the same courses and in other countries students could take various combinations of courses. This characterizes science instruction as received. These course-taking issues are treated extensively in Chapter 7. The reader is directed to other characteristics tabulated in Tables 2.6, 2.7, and 2.8 to obtain a rounded perspective on the way in which the schools organized their science teaching and facilities.

Teaching Load

It will be recalled from Table 2.7 that the average number of hours taught per week ranged from 14 to 24. These hours were then split between science teaching and other teaching and the average percent of all teaching devoted to science teaching calculated. This ranged from 43 percent in Italy (Grade 8) to 91 percent in Korea and Thailand. If 80 percent is used as a cut-off point to designate specialization in science then the teachers in Canada (Fr.), Ghana, Japan, Korea, the Netherlands, Nigeria, Philippines, Thailand, and Zimbabwe could be regarded as specializing in science teaching. In Australia, Finland, Israel, Papua New Guinea, Poland, Singapore, and the United States teachers taught science for between 70 and 80 percent of the time. However, in Hong Kong, Hungary, Italy (Grade 9), and Sweden the percentage devoted to science teaching of all teaching was, on average, between 60 and 70 percent.

It might be assumed that the higher the percentage of teachers' time allocated to the teaching of science then the more likely it is that the teachers are specialist science teachers. If this is so, then those countries where less than 80 percent of the time is spent teaching science might wish to reconsider their policies of the time allocation for teachers. However, it is possible that those teachers who teach both science and mathematics may be better science teachers because of this.

From the data collected, it was possible to derive which courses were taught by the teachers in which grade levels. A course is a general science course, or biology course, or chemistry course, or physics course, or earth science course.

Table 6.1 presents the percentages of all teachers in the sample in a country teaching courses at each particular grade level. Thus, for example, in Australia 21 percent of the teachers taught courses at Grade 7, 61 percent at Grade 8, 66 percent at Grade 9, 63 percent at Grade 10, 46 percent at Grade 11, and 35 percent at Grade 12. Because any one teacher teaches across several grades, the total adds up to more than 100. It may seem surprising that the box for the focal grade(s) involved in Population 2 does not have 100 percent. However, some countries asked all teachers teaching at that cycle of education to reply to the teacher questionnaire. If only the teachers teaching a specific class in the focal grade had been asked to answer the questionnaire, then 100 percent would have been in the focal grade. This is the case for Thailand.

TABLE 6.1 Percent of sections at each grade level, taught by focal grade teachers, by country (focal grade for each country boxed)

Country	Grade Level								
	4	5	6	7	8	9	10	11	12	13
Australia	0	0	0	21	61	66	63	46	35	0
Canada (Fr)	0	0	0	10	5	98	16	18	0	0
England	0	0	0	44	52	97	80	73	37	27
Finland	0	0	0	66	98	64	7	1	1	0
Ghana	1	0	0	20	31	55	62	60	20	25
Hong Kong	0	0	0	35	97	39	21	19	6	0
Hungary	9	26	47	64	93	0	0	0	0	0
Italy (Grade 8)	1	0	88	90	94	0	0	0	0	0
Italy (Grade 9)	0	0	0	0	0	92	80	43	31	24
Japan	0	0	0	30	40	78	0	0	0	0
Korea	0	0	0	57	50	51	0	0	0	0
Nigeria	0	0	0	6	8	28	51	15	0	0
Philippines	0	0	0	15	19	97	10	0	0	0
Poland	43	45	72	80	70	0	0	0	0	0
Singapore	0	0	0	7	5	97	54	0	0	0
Thailand	0	0	0	22	35	100	16	8	4	0
Zimbabwe	0	0	0	0	37	93	35	18	9	6

From Table 6.1 it also possible to see the grade-span of the schools involved in the sample. For example, in Australia the schools involved generally covered Grades 7 to 12, in Canada (Fr.) Grades 7 to 11, in England Grades 7 to 13 and so on.

It was also possible to produce four categories of grade-spans taught by each individual teacher. These categories were:
1. Teaching at the focal grade and also grades below that level.
2. Teaching at the focal grade only.
3. Teaching at the focal grade and at grades below and at grades above the focal level.
4. Teaching at the focal grade and at grades above that level.

Table 6.2 presents these data. In this case the figures are the percentage figures of teachers teaching within each grade range and add up to 100 percent.

TABLE 6.2 *Percentage of teachers with specific grade-span responsibilities, by country (categories centred on focal grade)*

Country	School Grade Spans	Teacher Grade-Span Category Focal Grade Level:			
		At and Below	At	Below to Above	At to Above
Australia	All	16	4	52	27
Canada (Fr.)	All	11	58	2	29
England	M and U	5	9	59	27
Finland	Middle	18	14	49	19
Ghana	M and U	16	7	21	56
Hong Kong	M and U	16	29	20	35
Hungary	Lower	73	27	0	0
Italy (Grade 8)	Middle	96	4	0	0
Italy (Grade 9)	Upper	0	10	0	90
Japan	Middle	56	44	0	0
Korea	Middle	81	19	0	0
Nigeria	M and U	37	40	13	10
Philippines	'Upper'	27	63	2	9
Poland	Lower	95	5	1	0
Singapore	All	8	38	4	51
Thailand	All	36	43	4	17
Zimbabwe	Middle	23	43	16	18

At the same time, the school cycles where the teaching primarily takes place have been categorized in the first column of the table. These categories are

- Lower schools only - *'Lower'*
- Middle schools only - *'Middle'*
- Upper schools only (including some schools with full grade-spans) - *'Upper'*
- Middle and Upper schools only (including some schools with full grade-spans) - *'M and U'*
- All school types - *'All'*.

Of countries with schools of all grade-span types represented (lower, middle, upper, all), all have teachers who teach in all focal grade-span categories. For countries with only middle and upper schools, all teacher grade-span focal categories are also filled. The only countries with lower schools only are Hungary and Poland and their teachers are consistently represented as teaching only at or below the focal grade. The countries with middle schools only - Italy (Grade 8), Japan, and Korea - all have homogeneous grade-spans which terminate at the focal grade level and the teachers consistently indicate this by only reporting teaching at or below the focal grade.

Countries with upper schools only are Italy (Grade 9) and the Philippines. Italy (Grade 9) has primarily Grades 9-13 schools and its focal grade is 9. Its teachers consistently report teaching either solely at Grade 9 or at and above this grade level. The Philippines has primarily Grades 7-10 *upper* schools as Grade 10 is the exit grade for secondary school. Since its focal grade is 9 it consistently reports all teaching grade categories.

Which Subjects Do Teachers Teach?

Table 6.3 presents the percentage of teachers teaching the various subjects. For this table, earth science courses have been incorporated into general science courses. [Finland and the Philippines are excluded because of data anomalies.]

TABLE 6.3 Percent of teachers of science teaching various subjects, by country at Population 3

Percent of Teachers of Science Teaching Various Subjects
All Grades

Country	General or Earth Science	Biology	Chemistry	Physics	Other
Australia	95	26	20	18	5
Canada (Fr.)	8	76	8	31	10
England	66	42	37	40	16
Ghana	50	34	31	28	5
Hong Kong	96	17	17	8	0
Hungary	40	5	28	27	28
Italy (Grade 8)	100	0	0	0	0
Italy (Grade 9)	61	6	16	31	41
Japan	100	0	0	0	0
Korea	100	0	0	0	0
Nigeria	5	29	33	27	16
Poland	37	32	26	32	0
Singapore	55	19	23	20	29
Thailand	100	5	2	5	7
Zimbabwe	92	13	6	5	5

Percent of Teachers of Science Teaching Various Subjects
Focal Grade

Country	General or Earth Science	Biology	Chemistry	Physics	Other
Australia	94	3	3	3	5
Canada (Fr.)	7	77	0	26	2
England	18	35	32	34	4
Ghana	47	18	22	22	4
Hong Kong	99	0	1	2	0
Hungary	28	4	28	30	25
Italy (Grade 8)	100	0	0	0	0
Italy (Grade 9)	36	8	4	28	33
Japan	100	0	0	0	0
Korea	100	0	0	0	0
Nigeria	4	33	35	31	6
Poland	21	23	33	40	0
Singapore	47	19	23	20	27
Thailand	100	0	0	0	1
Zimbabwe	94	2	2	2	1

The predominant science courses taught in most countries are general or earth science only. In seven out of the 15 countries analyzed, more than 90 percent of teachers teach general science courses. In a further four countries, more than 50 percent teach such courses.

It is also possible to classify the teacher data in a different way:
1. those teaching only general science (including earth sciences);
2. those teaching a single specialized subject only (e.g., physics);
3. those teaching at least two specialized science subjects; and
4. those teaching both general and a single specialized science subject.

These data, representing teaching responsibilities at all grades, are presented in Table 6.4.

TABLE 6.4 Percent of teachers of science with general and specialized science teaching loads, by country

Country	General or Earth Science Only	Specialized Science Only		Both General and Specialized Science
		One Subject	Two or more	
Australia	41	3	1	55
Canada (Fr.)	2	65	26	7
England	2	24	9	65
Ghana	19	40	10	32
Hong Kong	63	4	0	35
Hungary	19	53	6	21
Italy (Grade 8)	100	0	0	0
Italy (Grade 9)	15	36	3	47
Japan	100	0	0	0
Korea	100	0	0	0
Nigeria	2	77	17	3
Poland	26	48	15	11
Singapore	27	33	12	29
Thailand	83	0	0	17
Zimbabwe	74	3	2	19

The countries can be grouped into five homogeneous categories on the basis of the distribution of teaching specialties within the country. [It should be noted that in some countries, e.g., Hungary or Poland, which have fixed curricula for students, teaching responsibilities can be varied in mounting the required patterns.] These categories fall into a rough order from countries with specialized teachers only to countries which have only general science teachers. These categories and the countries belonging to them are as follows and confirm the teaching load patterns above.

Most teachers teaching one specific science subject only:	Canada (Fr.) Nigeria
Most teachers teaching specific subjects but some also teaching general science:	England
A variety of patterns - general science only, specific subject only, and mixed:	Italy (Grade 9) Ghana Hungary Poland Singapore
Most teachers teaching general science, but some also teaching specific subjects:	Australia Hong Kong Thailand Zimbabwe
All teachers teaching general science only:	Italy (Grade 8) Japan Korea

Conclusions

This chapter has examined the Population 2 teachers in terms of teaching load, grades taught and subjects taught.

The overall number of hours per week taught altogether ranged from 14 (Nigeria) to 24 (USA). The number of hours teaching science courses ranged from 7.3 in Italy (Grade 8) to 18 in Korea. The number of class science courses (not individual lessons) taught ranged from two in Italy (Grade 8) to seven in England. The typical amount of time per course was between two and three hours per week. Science teachers also taught some non-science subjects. The amount of time devoted to the teaching of non-science courses was, on average, 4.3 hours or 23 percent and the percent ranged from nine percent in Korea and Thailand to 57 percent in Italy (Grade 8).

The number of grades taught was, in part, a function of the way in which the school system was structured. Population 2 was defined as the grade in which most 14-year-olds were enrolled. In Hungary and Poland this was Grade 8 which was the end of the cycle of basic schooling. The science teachers in these two countries typically covered the range of Grade 4 to Grade 8. In other systems it was either Grades 7 to 9 where there was a separate cycle of schooling for these grades or, where there was a school covering Grades 7 to 12 or 13, then this was the gamut of grades taught.

In terms of science subjects taught, the most common practice was that of teachers teaching general or integrated science or earth science. The specialized teaching of biology, chemistry, or physics was less common. In Japan and Korea only the subject 'science' was taught. This was, for the main part also true in Hong Kong, Italy (Grade 8), and Thailand. In England, and less commonly Australia, teachers often taught a mixture of general science and specialized sciences. In Canada (Fr.), Finland, Ghana, Hungary, Nigeria, and Poland the teachers tended to teach only courses in biology, chemistry, or physics.

In the next chapter in this volume (Chapter 7), the analyses consider whether the students achieve more when they are in schools which offer biology, chemistry, and physics as separate, specialist, and compulsory courses. Thus, the findings in this chapter need to be read with the findings of Chapter 7 in mind.

7

The Science Curriculum and Achievement

All schools at the Population 2 level have a science curriculum. They offer science courses and, as was seen in the last chapter, there are general science teachers and specialist teachers of biology, chemistry, and physics. Some schools have general science courses and also separate science subject courses. Some have only some of the separate science courses. And, in some schools, courses are compulsory and, in others, they are optional.

This chapter examines what schools offered as science curricula, which subjects were compulsory and the achievement of students on the IEA science tests according to what the school curriculum was and what it required.

It was possible to use the information provided by the students, the teachers, and the schools in order to extract data on what schools offered and, where nearly 100 percent of the age group were enrolled in the Population 2 grade level, and which subjects schools made compulsory. 'Compulsory courses' in some school systems' jargon is the same as 'required courses'.

Average Offerings and Requirements

Table 7.1 presents the percentages of schools offering general science, earth science, biology, chemistry, and physics at the target grade for Population 2. General Science is defined as instruction in which specialty science subjects are

TABLE 7.1 Percent of schools with offerings (by subject and country) at Target grade for Population 2

Country	Target Grade	General	Earth Science	Biology	Chemistry	Physics
Australia	8,9,10	97.6	1.9	4.8	5.7	5.7
Canada (Eng.)	9	90.1	1.2	7.4	0.6	3.7
Canada (Fr.)	9	41.8	8.8	81.3	1.1	18.7
England	9	39.7	0.7	84.6	82.4	83.1
Finland	8	0.0	100.0	100.0	100.0	100.0
Hong Kong	8	97.2	0.0	1.9	2.8	2.8
Hungary	8	0.0	100.0	100.0	100.0	100.0
Italy (Grade 8)	8	88.4	23.2	16.6	19.9	22.7
Italy (Grade 9)	9	61.0	34.1	22.0	19.5	63.4
Japan	9	100.0	0.0	0.0	0.0	0.0
Korea	9	100.0	0.0	0.0	0.0	0.0
Netherlands	9	2.1	71.1	66.3	56.7	78.6
Nigeria	10	4.2	0.0	98.6	97.2	98.6
Papua New Guinea	10	100.0	0.0	0.0	0.0	0.0
Philippines	9	0.4	1.6	5.2	94.4	0.4
Poland	8	100.0	100.0	100.0	100.0	100.0
Singapore	9	0.0	0.0	76.5	60.2	62.7
Thailand	9	100.0	0.0	0.0	0.0	0.0
U.S.A.	9	75.3	37.0	45.2	4.1	0.0
Average		57.8	25.2	42.7	39.2	39.0

mixed or integrated rather than taught as separate subjects. The percentages average over the varied patterns containing these subject offerings.

It should be noted that, in contradistinction to Chapter 6, Earth Science is here separated from General Science. Earth Science in Finland, Hungary and Poland represents Social and Economic Geography. General Science in Poland is anomalous, given other offerings.

Upon examination of the table, it can be seen that Finland and Hungary both display the same offering pattern. They offer geography, biology, chemistry and physics and not general science at the focal grade in every school. Japan, Korea, and Thailand, on the other hand, offer only general science and no specialized subjects in every school. [It should be noted that Japan provides two mixed areas: Earth Science with Biology, and Physics and Chemistry.] Poland offers all five of the subject areas in each school (again with Geography rather than Earth Science). Other countries offer a mixture of patterns with no subject offered in either 100 percent or zero percent of the schools. The exceptions to this are Hong Kong and Nigeria (no earth science offered at the focal grade), and United States (Phase 1) where no physics is separately offered. (However, the United States offers Physical Science at Grade 9 in some schools.)

In this table, countries also conform to predominant patterns if the criterion of 100 and zero percent offerings in a subject area is relaxed. Australia offers mostly general science as a subject (98%) and rarely earth science (2%) and only occasionally biology, chemistry, and physics (5 to 6%) as separate subjects at the focal grade. Canada (Eng.) and Hong Kong also have similar patterns.

Canada (Fr.) only offers general science in 42 percent of its schools, but offers biology in 82 percent. To some degree this part of Canada is similar to Italy in that offerings are spread among all five types with a substantial offering in general science, but in Italy (Grade 8) no specialty science predominates as does biology in Canada (Fr.), and in Italy (Grade 9) there is a substantial offering of physics (63%) at the focal grade.

The Netherlands has substantial offerings in earth science, biology, chemistry, and physics (57-79%), but only a small number of schools which offer general science (2%). Singapore is somewhat similar with 60 to 77 percent in biology, chemistry, and physics, but no offerings in either general or earth science. Nigeria predominantly offers biology, chemistry, and physics (97 to 99%) with little general science (4%) and no earth science. The Philippines offers mostly chemistry (94%) with small numbers of schools offering biology (6%); the other subjects are each offered by less than two percent of Philippine schools. The United States offers mostly general science (75%), biology (45%), and earth science (37%). A few United States schools also offer chemistry (4%) at the focal grade, but none offer physics.

A compulsory or required subject is one in which all students in a grade participate. This encompasses both specific subject area requirements and required subject options, e.g., students being required to take **either** chemistry or physics. The information was obtained by matching course or subject enrollments with total grade enrollments for each school.

TABLE 7.2 Percent of schools with science subject requirements (by country) at Population 2 level

Country	Missing Data (% total)	None	By Subject						Combined		Total Schools (Freq.)
			General	Earth	Biology	Chemistry	Physics	BCP	GEBCP	
Australia	10	16.3	81.8	1.0	2.9	2.4	2.9	1.9	0.0	233
Canada (Eng.)	22	26.5	65.4	0.6	4.3	0.6	3.7	0.6	0.0	209
Canada (Fr.)	10	28.6	18.7	4.4	59.3	1.1	4.4	1.1	0.0	101
England	7	11.0	32.4	0.0	75.0	68.4	69.9	67.6	0.0	147
Finland	1	0.0	0.0	100.0	100.0	100.0	100.0	100.0	0.0	90
Hong Kong	3	3.7	93.5	0.0	1.9	2.8	2.8	1.9	0.0	110
Hungary	2	0.0	0.0	100.0	100.0	100.0	100.0	100.0	0.0	99
Italy (Grade 8)	18	4.4	85.6	21.5	14.9	18.8	20.4	10.5	7.2	222
Italy (Grade 9)	42	9.8	58.5	31.7	14.6	12.2	53.7	0.0	0.0	71
Japan	0	0.0	100.0	0.0	0.0	0.0	0.0	0.0	0.0	199
Korea	1	0.5	99.5	0.0	0.0	0.0	0.0	0.0	0.0	189
Netherlands	17	23.0	0.5	48.7	34.2	41.7	48.1	13.9	0.0	224
Nigeria	13	31.0	2.8	0.0	69.0	39.4	38.0	38.0	0.0	82
Papua New Guinea	5	0.0	100.0	0.0	0.0	0.0	0.0	0.0	0.0	78
Philippines	8	10.0	0.0	1.2	4.0	85.5	0.0	0.0	0.0	272
Poland	1	0.0	100.0	100.0	100.0	100.0	100.0	100.0	100.0	201
Singapore	10	98.2	0.0	0.0	1.8	1.8	0.6	0.6	0.0	185
Thailand	4	0.0	0.0	0.0	0.0	0.0	0.0	0.0	0.0	96
U.S.A.	17	60.3	28.8	11.0	5.5	0.0	0.0	0.0	0.0	88
Average/Total		17.0	50.9	22.1	30.9	30.3	28.7	23.0	5.6	2,896

In analyzing school requirements it is clear that the possibility of having no requirements arises. In Table 7.2, a variety of requirements which can be established in schools offering the requisite courses is exhibited. The first column gives the percentage of schools in each country for which the enrollment data were not available. The remaining columns relate to requirements. The percentages displayed there represent the percentage of schools in each country, using the number non-missing enrollment data as the base, requiring a subject or a combination of subjects.

The second column displays the percentage of schools in each country with no specific science subject requirements. It can be seen that the distribution of this percentage is highly skewed over countries. In six countries all of the sampled schools have requirements. These are Finland, Hungary, Japan, Papua New Guinea, Poland, and Thailand. For all of the other countries but two, the percentages of schools without requirements ranges from one to 31 (Nigeria). After this there is a substantial gap until the United States (60%) and Singapore (98%). Thus, most countries have science subject requirements for students in the Population 2 focal grade (Table 7.1) in the vast majority of their schools. Only the United States and Singapore are exceptions to this general rule.

TABLE 7.3 Most frequent offering and enrollment patterns (by country) at Population 2 level

Country	Pattern of subjects		Percent Total Schools with pattern
	Offered	Required	
Australia	G	G	78.4
Canada (Eng.)	G	G	64.8
Canada (Fr.)	B	B	27.5
England	BCP	BCP	50.0
Finland*	EBCP	EBCP	100.0
Hong Kong	G	G	93.5
Hungary*	EBCP	EBCP	100.0
Italy (Grade 8)	G	G	69.6
Italy (Grade 9)	G	G	22.0
Japan	G	G	100.0
Korea	G	G	100.0
Netherlands	ECP	ECP	11.2
Nigeria	BCP	BCP	33.8
Papua New Guinea	G	G	100.0
Philippines	C	C	83.9
Poland*	GEBCP	GEBCP	100.0
Singapore	B	none	37.3
Thailand	G	G	100.0
U.S.A.	G	G	21.9

G - General Science   P - Physics   E - Earth Science
C - Chemistry   B - Biology   (*Geography)

System Pattern

It was possible to list the patterns of what was offered in the science curriculum and what was compulsory or required.

In extracting this information it was possible to characterize each school with respect to (a) subjects offered (G-general science, E-earth science, B-biology, C-chemistry, P-physics) and whether particular subjects or combinations of them were compulsory. Among all schools in all countries for which there were enrollment data, the most frequent pattern was 'general science only, required'. This was the only focal grade pattern for 42 percent of all schools. In four percent 'general science, not required' was the pattern. In the remaining schools (54%) either a specific single (non-general) subject or some combination of subjects were offered. All in all, 41 percent of the latter schools had some form of requirement associated with these offerings.

Table 7.3 displays the most frequent offering and requirement pattern for each country. It is notable that, except for Singapore, all countries have an offering which is also compulsory (required) as the most frequent. The required general science offering is uniform in four countries (Japan, Korea, Papua New Guinea, Thailand) and predominant in four (Australia, Canada (Eng.), Hong Kong, Italy (Grade 8)). It is also the most frequent within a varied set of patterns in two (Italy (Grade 9), United States (Phase 1)). However, several other patterns which were listed as composite requirements - GEBCP and EBCP - may actually refer to integrated general science offerings and requirements.

Appendix H presents all patterns with frequencies greater than one. It presents the number of schools (for which data were available) which offered different courses and which required them. A perusal of Appendix H shows that other patterns which emerged are **biology, chemistry, and physics - all required** (England), and **chemistry only - compulsory** (Philippines). There is considerable variation in the other six countries in terms of the most frequent offering and requirement patterns encompassing 50 percent of those countries' schools. Canada (Fr.) and Italy (Grade 9) are mixed between general science and specialty subjects - all required. Canada (Fr.) has five patterns mixing general science with biology; in Italy (Grade 9) (10 patterns) general and non-biological sciences form the patterns - again all required; in the United States (three patterns) general science is predominant, sometimes mixed with biology, and sometimes required. The Netherlands (seven patterns), Nigeria (two patterns), and Singapore (two patterns) have specialty courses only as frequent patterns, but Nigeria requires its most frequent patterns, the Netherlands requires most of them, and Singapore requires none of them.

From examining the grade levels preceding the target population grades in an analysis not reported here, it was apparent that in 12 of the 19 countries the distributions of offerings and requirements were substantially the same as those in the focal grades. In the remaining seven, however, there were significant differences.

In *Canada (Fr.)*, the mixed general-specialty pattern at the focal grade (9) changes somewhat from a greater proportion of schools with required general science.

In *England*, required general science is the norm in the earlier grade (8) rather than the mixture of specialties that characterize the focal grade (9). In the majority of the English schools the difference between Grades 8 and 9 seem to signify a system-wide, rather than a school-specific, grade-span shift.

In *Finland*, the predominant pattern at the focal grade (8) was required biology, chemistry, and physics. At the earlier grade (7), it was the same except that chemistry was not a part of the offering-requirement pattern. Again, this seems to be a system-wide, within-school shift in that most of the Finnish schools begin at Grade 7.

In *Italy (Grade 9)*, the mixed pattern at the focal grade (9) characterizes the initial grade in all schools. The previous grade is the focal grade for Italy (Grade 8), and there the predominant pattern is required general science. The grade-span shift in Italy corresponds to a system-wide between-school shift in curriculum that accompanies the system shift to differentiated schools.

In the *Philippines*, required chemistry is the predominant pattern at the focal grade (9), while required biology is predominant in the preceding grade. This is a system-wide within-school shift in curriculum.

In *Singapore*, the non-required, mixed, specialty-only pattern at the focal grade (9), is preceded by a clear pattern of required general science. The transition from Grade 8 to Grade 9 in Singapore clearly is a double shift - general to specialty and required to non-required. Since the predominant grade-span pattern in Singapore is 7 to 10, this is again a system-wide, within-school shift.

In the *United States*, the shift is gradual. In the focal grade (9), the pattern is mostly general - some required, some not - with some admixture of specialty courses. In the previous grade, the pattern is all general - with some requirements. This would seem to be a mixture of patterns. In the United States, Grade 8 is not usually in a high school (9 to 12 or 10 to 12), but Grade 9 could be either in a middle school (7 to 9) or a high school (9 to 12).

Patterns, Opportunity to Learn, and Achievement

It is clear that the different curricular patterns (which courses are offered or required) may well affect the opportunity to learn (OTL) offered to the students and their achievement. Table 7.4 presents OTL and achievement data for twenty-five course patterns. The pattern code is made up of ten values. The first five indicate if the subjects - general science, earth science, biology, chemistry, and physics - are offered and the second five whether they are compulsory or not. Thus, the first pattern is **P P** which indicates that physics is offered and compulsory; the second pattern (C -) indicates that chemistry is offered but not compulsory. The last pattern (**GEBCP GEBCP**) indicates that all five subjects are offered and compulsory.

Opportunity to Learn data, where available, are presented for each of the four test areas and exhibit large variations across the curricular patterns. For example, in biology the average percentage of test items covered ranges from 23 percent to as high as 73 percent. The lowest value is for those schools (across all countries) in which the curricular pattern is **general science only is offered but not required**. The highest value is for the pattern in which **earth science, biology, chemistry and physics** are all **offered** and each of these is also **required**.

In earth science the average percentage of items covered (OTL) varies from 20 to 60 percent; in chemistry the range is 30 to 51 percent and in physics it varies

The Science Curriculum and Achievement

between 25 and 70 percent. The smaller range for chemistry indicates a greater degree of homogeneity across schools in opportunities for learning that area.

The implications of such large variations are profound. Obviously students attending schools in these 24 systems are subject to drastically different educational opportunities in science at least as defined by the tests used in the Second IEA Science Study.

Do the opportunities to learn vary across patterns only or do they also vary across the four test areas? Is the relationship between course offerings and achievement more complicated in that it is different for each test area (an interaction between the patterns and tests)? To answer these questions the data were subjected to a repeated measures analysis of variance.

TABLE 7.4 *Twenty-five course patterns, and the achievement and OTL means*

Pattern Offered	Pattern Required	Earth Science OTL	Earth Science Achv.	Biology OTL	Biology Achv.	Chemistry OTL	Chemistry Achv.	Physics OTL	Physics Achv.
P	P		67.9		52.5		56.0		68.6
C	-	34.4	44.4	50.2	31.7	40.7	39.0	31.5	37.7
C	C	45.0	46.0	58.5	34.8	50.8	42.2	43.6	40.3
CP	-	25.2	61.0	42.1	46.5	44.5	52.5	48.3	59.1
C or P	-	19.6	61.7	35.1	48.9	29.7	52.5	36.2	61.2
B	-	26.7	59.5	38.5	45.9	35.7	54.0	39.7	57.2
B	B	44.4	55.5	58.7	42.8	42.9	53.3	46.7	51.7
BCP	-	34.9	51.5	29.8	38.9	32.6	48.5	38.7	53.5
BCP	BCP	38.7	57.3	48.6	46.8	44.4	53.3	51.4	55.3
B or CP	-	38.6	68.8	42.5	58.8	47.9	65.4	49.6	66.8
B or C or P	-	31.5	57.3	26.8	49.7	35.5	55.2	35.1	55.4
BC or P	B		43.5		38.2		50.1		44.6
E	E	33.3	66.3	26.1	49.6	42.9	59.2	25.0	58.1
ECP	ECP		75.6		70.0		67.5		73.7
EC	EC		63.5		48.5		59.0		60.6
EBCP	EBCP	60.1	71.2	73.3	59.8	63.4	64.0	69.7	66.2
EBCP	EBC		66.2		57.8		59.5		63.4
E or B	-		55.5		41.5		51.4		50.4
ECB or P	EC		72.2		61.9		62.7		67.2
G	-	25.9	65.9	23.2	51.0	35.7	60.8	33.3	61.1
G	G	51.8	65.9	52.3	51.6	44.6	59.4	51.5	61.6
G or B	-		63.3		46.4		58.7		56.7
G or B	B		63.5		51.2		59.6		59.5
G or BCP	G or BCP	37.6	58.8	35.4	49.1	41.8	52.7	43.9	58.0
GEBCP	GEBCP		64.7		56.2		59.9		56.2

The results of these analyses indicated that the effect of curricular pattern on achievement differed from one science subject area test to another. This is expected in that different patterns include different combinations of subjects. The non-uniformity in the effects is a validation of the analysis since it corresponds to the expectation. An index of the relative magnitude of this effect can be calculated from the analysis of variance. Using this computation, the magnitude of the variation (in standard deviation units) of the effect is over one-half (59 percent) of the size of the average effect, which is sizable. The results imply that the different patterns have considerably different effects on opportunity to learn depending upon the test area.

For example, in schools which only offer biology at the Population 2 level and make it mandatory, the average percentage of the biology items covered is 59 percent while coverage of the items in the other three tests is only around 45 percent. By contrast, the average percent of biology items covered by schools in which the curriculum offering is either a course in biology or two courses (chemistry and physics) taken together (neither of which is required) is 43 percent while coverage of the physics and chemistry items is 50 percent.

From an analysis undertaken but not presented in tabular form here, a statistically significant variation in the effect of curricular pattern across achievement subject area was detected but this heterogeneity was of the order of five times smaller than the average effects of patterns on achievement. This is a fraction of the size of the corresponding effect on OTL. The likely explanation is that Opportunity to Learn, as measured, is more valid for the focal and closely prior adjacent grades than for early grades. Achievement, however, is likely to be highly cumulative from grade to grade.

The nature of the non-uniformity in effect on achievement is clear upon closer examination of Table 7.4. There are patterns for which achievement is higher in one area of testing than in the others while there are other patterns in which the achievement profile is different in the separate subjects. However, the differences are not very large and, where present, reflect only subtle distinctions. For example, the average achievement in schools which require only physics at the Population 2 level is among the highest in physics but is only average in biology; on the other hand, in schools in which earth science, biology, chemistry and physics are required, the average achievement is higher in biology and lower in physics than for the previous pattern.

Several analyses were undertaken in which the effects of requirements were formally analyzed using linear models. These were performed by country. In several cases these analyses revealed strong positive effects of required science on achievement. These results support the obvious proposition that requirements promote exposure and participation which, in turn, increase learning. But more generally one is struck with the more impressive observation that the profile of achievement is relatively parallel across subject tests but different among the curricular patterns. Note for example that in schools in which the students are required to take all subjects except general science one finds the highest achievement levels across all four subject tests.

The general impression one is left with is that the pattern of course offerings and requirements characteristic of different schools is strongly related to achievement. It is also noted that although these differences vary in a statistical way across subject test areas, there is little practical significance to these differences and that what is clearly more important are the general differences one finds across all four curricular areas.

The implication is that the curricular patterns reflect fundamental differences in the organization of the curriculum that transcend the Population 2 school year and as a result affect science achievement in a cumulative way. This would explain the apparent undifferentiated effects of course offerings and requirements on achievement.

The social class status of the school is related both to the type of course offerings and requirements found in that school and to the general achievement level of its students. A question that remains is whether the differences in achievement across patterns reflect only the social class differences associated

The Science Curriculum and Achievement

with the patterns or if after accounting for the achievement differences that result from the social class differences there are still major achievement differences related to curricular patterns.

A further analysis (repeated measures ANOVA) was run adjusting the scores by taking out the effects of social class. The results shown in Table 7.3 did not change significantly. In other words, the way in which the curriculum is organized in a school is critical in terms of what type of science curriculum students are offered and how they achieve.

In summary, the pattern of course offering and requirements is strongly related to achievement in the pooled sample of schools from participating countries and in several of the specific countries. This relationship also holds after control for social class. One of the components of this relationship relates to requirements. Schools requiring specific subjects are likely to have higher achievement in those subjects than schools which do not require student participation at the focal grade.

Conclusions

The types of science courses offered and required in all of the schools in the 19 countries at Population 2 for which data were available were examined.

The percentage of schools (N=2,610) offering and requiring each of the subjects were:

	General	Earth Science (Geography)	Biology	Chemistry	Physics
Offering	59	24	38	41	35
Requiring	53	21	28	33	26

Over all countries 42 percent of schools required the Population 2 student to take general or integrated science courses only.

The overall patterns are summarized below:

General science only - compulsory:	Japan, Korea Papua New Guinea, Thailand
Earth Science, Biology, Chemistry, Physics - compulsory:	Finland, Hungary
All subjects - compulsory:	Poland
General Science only - offered:	Australia, Canada (Eng.) Hong Kong, Italy (Grade 8)
Chemistry only - offered:	Philippines
Biology, Chemistry, Physics - offered:	England, Nigeria
Scattered patterns:	Canada (Fr.), Italy (Grade 9) Netherlands, Singapore United States

The most common (more than 50 percent of a country's schools) pattern of compulsory subjects was general or integrated science.

General or Integrated Science	Australia, Canada (Eng.) Hong Kong, Italy (Grade 8) Japan, Korea Papua New Guinea, Thailand
Biology, Chemistry, Physics:	England
Earth Science, Biology, Chemistry, Physics:	Finland, Hungary
Chemistry only:	Philippines
General Science, Biology, Chemistry, Physics:	Poland

Although definitive evidence is not available from the analyses performed, the data lend support to two more speculative conclusions:

1. Where schools offer separate science subjects the students, in general, are offered more to learn. In other words, the opportunity to learn is greater in the separate subject offerings than is the case in general and integrated science courses.

2. Schools having compulsory separate courses in biology, chemistry, and physics have higher achievement (even when the social class background of students is taken into account) than any of the other patterns of science courses.

8

Influences on Student Achievement

In the previous chapters the relationship of selected variables to student or school achievement has been examined. This chapter attempts to discuss the relationship of many more variables to achievement.

The methodological framework used was latent variable path analysis and the particular form of analysis used was partial least squares (PLS). The first step was to cluster conceptually similar measures into constructs. The second step was to examine the relationships among the constructs. Using Population 2 a small analysis was undertaken on seven countries. These were Canada (Fr.), China, Hong Kong, Korea, the Philippines, Poland, and Singapore. These countries were selected because the outcomes of similar analyses for Australia, England, Finland, Hungary, Italy, Japan, the Netherlands, Sweden, Thailand, and the United States have been reported in Volume 3 of this series (Keeves, 1991) and the other countries proved problematic due either to high missing data rates or because no teacher or school data were collected.

By combining the correlation matrices (with each country having a weight of 1.0) of all variables in each of the seven countries an overall analysis was run and the results reported in Figure 8.1.

This figure will be used to explain how the analyses work. In Figure 8.1 there are a series of boxes each of which has a title. The box at the top left of the figure is entitled 'Socio-Economic Status of Home' (SES). This is a construct made up of family size, father's education, mother's education, and father's occupation. The figures next to each variable indicate the relative importance of a variable in the construct. In this case father's occupation is the most important part of the construct with a loading of .90. Then come father's education (.65), mother's education (.60), and family size (-.39). This indicates for the seven countries that high SES homes have fathers with a high status job, relatively good education and small families. Low SES homes have the opposite characteristics: fathers with low status jobs, parents with poor education and large families. The 'Literacy of Home' construct contains three variables of which the most important is 'Books in Home'.

Each construct can be examined for the conceptual similarity of the variables in the construct and for the relative weight they carry in the construct.

The second feature of Figure 8.1 is that arrows have been drawn from one construct to another to indicate possible causal relationships. There are five arrows from the SES construct. In this model it was posited by the research group that the SES would influence the constructs 'View of and Interest in Science', 'Science Achievement', 'Like School', 'Homework', and 'Literacy of Home'. In this case, the path coefficients are .10, .11, -.01, .15, and .43 respectively. In this study which is based on a probability sample in each country, it was decided, after consideration of the accuracy of the estimates, that .04 and below indicates no effect, .05 to .09 a weak effect, .10 to .19 a clear effect, and .20 or more a strong effect. Thus, SES is referred to have no effect on 'Like School' but to have an effect on 'View of and Interest in Science', 'Science Achievement', and 'Homework' and to have a strong effect on 'Literacy of Home'.

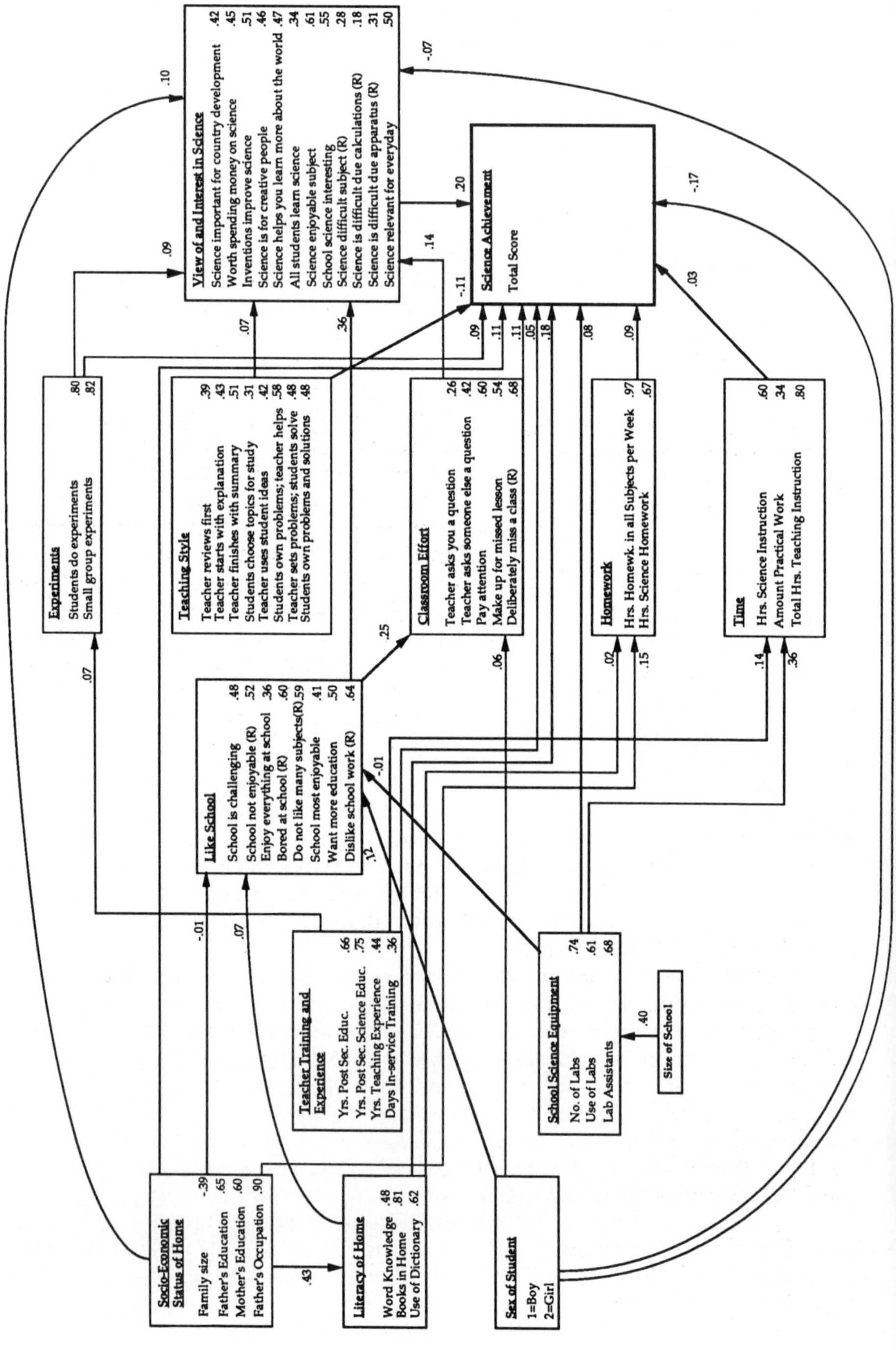

FIG. 8.1 PLS-model of pooled within and between construct effects (Population 2)

Influences on Student Achievement

Another way of disentangling the figures is to work backwards or from achievement to its sources. Working anti-clockwise, the constructs influencing science achievement and their coefficients are:

Construct	Path Coefficient	Influence
View of and Interest in Science	.20	strong positive effect
Teaching Style	-.11	negative effect
Experiments	.09	weak positive effect
Socio-Economic Status of Home	.11	positive effect
Classroom Effort	.11	positive effect
Teacher Training and Experience	.05	weak positive effect
Literacy of Home	.18	clear positive effect
Homework	.09	weak positive effect
Time	.03	no effect
Sex of Student	-.17	boys do better, clear effect

It is common in the presentation of path analysis to remove paths with a coefficient of less than 0.08. In Figure 8.1, however, all path coefficients were left in. The reader will also see from Appendix I that there was some variation of effects for particular variables from country to country.

It may be of interest to some readers to know more about the rationale of the analyses. In all, a total of 56 measures were used in the analyses. These measures were arrayed in 13 clusters, two of which, 'Sex of Student', and 'Science Achievement', contained only one measure each. The largest clusters contained 12 ('View of and Interest in Science'), 8 ('Like School'), and 8 measures ('Teaching Style') each. In the framework of the formal model, each cluster is identified with a latent variable or *construct*.

By organizing the 56 measures into these clusters, the total number of interrelations among them is reduced. Without clustering, there would be 1,540 relations among the measures. Clustering reduces this to 78 between-cluster and 43 within-cluster relations, for a total of 121. This is a 92 percent reduction in the number of relations needed to describe the data. This reduction significantly reduces the cognitive task of interpretation. The pitfall in clustering is that if the groups are not formed in a fashion which reflects meaningful homogeneity within the clusters - both conceptually and empirically - the cluster relation will not be meaningfully interpretable.

Once the clusters are specified, a second step is required. The intent of this step is to constrain the relations between the clusters. This takes the form of designating which clusters are directly related to one another and which clusters are not. These direct relations are asymmetric in the sense that when one cluster is presumed to 'influence' another - i.e., if the second is an outcome of the first - then there can be no 'influence' of the second on the first. In addition, 'circular' relations are not allowed, i.e., if cluster A affects cluster B, and cluster B affects cluster C, then cluster C cannot affect cluster A. These constraints make the total set of relationships 'recursive.'

Depending on the ultimate style of interpretation, this process may be more critical for the ultimate results than the initial clustering. If the values of the relational coefficients are given 'causal' interpretations - i.e., the values are thought to allow calculation of amounts of change resulting from specific modifications of a causal factor - then the omission of an important relational factor can bias the estimates of the coefficients for the remaining factors. If, on the other hand, the results are considered tentative and subject to modification in interpretation by additional scientific and practical knowledge, then reasonable interpretations are possible. In particular, specific coefficients can be viewed as 'adjusted' for potentially spurious confounding relations when other factors which are included also affect common outcomes. Usually these relations are specified iteratively, in that larger sets of relations are initially assumed and then are deleted if they do not appear empirically. This process is further described below.

In the implementation of these analyses using partial least squares (PLS), a relatively complicated computer program is used which successively estimates a set of weights for each measure in each cluster and a set of relational coefficients connecting the clusters. The estimation procedure is repeated, iteratively, until stable estimates are obtained. Because of the complexity of the computation, it was decided to cross-validate the results for the Population 2 analyses by implementing an alternate, two-stage procedure. First, a principal component analysis was performed for each cluster. The first component extracted in each analysis was used as a cluster summary variable for inter-cluster analysis. This was performed as a regression analysis with an identical structure to that in the PLS analysis.

Comparisons were then made between the two with respect to two aspects: (a) within cluster and (b) between cluster coefficients. The values were virtually identical for all comparisons, implying the robustness of the computational procedure. The between-cluster coefficient values for both procedures are given in Appendix I. Given this outcome, the PLS procedures were used for interpretation. It should be noted that this test of the 'robustness' of the computing algorithm does not assess the robustness of the model specification, i.e., of the composition of the clusters themselves or of the recursive constraints imposed on the between-cluster relations.

Substantively, the between-cluster relations were specified in a straightforward fashion. Basically, it was postulated that the socio-economic status of the home would influence the literacy of the home, both of which would influence the students' liking school, the homework they do, their attitudes to science and their science achievement. Furthermore, that the school's science facilities would influence the student's liking school, the time devoted to learning science and the student's achievement. The reader may refer to Figure 8.1 for details on which constructs are expected to influence which other constructs. It will be noted, in particular, that OTL is not included as a construct. This was because it was more difficult to find meaningful OTL relations in analyses between pupils and classes than in between country ones.

Construct weights

As a first step in screening country differences in construct composition, the mean value for the weight of each measure over the seven Population 2 countries was computed. Each of these means was subtracted from each of the seven relevant country weights. The distribution of the resulting deviations over all the measures was then tabulated.

Influences on Student Achievement   129

Extreme values in this distribution were then scrutinized. Most of these values were idiosyncratic, i.e., they did not systematically characterize specific countries or question clusters. The exception to this was the construct of 'Teacher Style' for the Philippines. Each of the five measures which indicated that the teacher gave some authority to students in the teaching process had highly negative discrepancies. Reference to the original weights (Appendix I.1) showed that they also were negative. In fact, among all the weights in the question by country array for 'Teaching Style' only these Philippine weights were negative.

To confirm this finding further, a two-way analysis of the 'Teaching Style' weights was performed. In this analysis, the array of weights was decomposed into country effects, question effects, and residual effects. This analysis was similar to those used in analyzing the OTL-achievement relation, above. This analysis confirmed the structural differences between the Philippine construct and the construct found for the other six countries. It seems that for Philippine respondents, the teacher style construct is really two: (1) a 'systematic teaching' component represented by the first three questions, and (2) a 'student orientation' component represented by the last five. These two components are somewhat negatively correlated in the Philippines. The 'Teaching Style' construct variable for the Philippines is therefore an arbitrary composite of these two components. The meaning of this construct is not the same as that for the other countries. This difference is also confirmed in the relational coefficients linking the cluster constructs. The direct effect of 'Teaching Style' on achievement is negative for all countries except Korea (.05) and the Philippines (.14). In fact, the Philippine value is most discrepant from the value from the pooled analysis.

Relational coefficients

In order to assess whether the overall results, presented above, were also satisfactory in characterizing each of the seven Population 2 countries, a two-way analysis of the relational coefficients linking the cluster constructs to achievement was undertaken. This analysis showed clear differences in the magnitudes of the coefficients pertaining to different clusters, but no overall differences between countries. This implies that, if the 'average' values of the construct effect match each of the corresponding values for every country, that the overall coefficients satisfactorily characterize each country.

To test this explicitly, the discrepancies between the overall values and the country specific values were examined. It was found that there were only five coefficient values which were not close to the overall values. One of these was the aforementioned value relating 'Teaching Style' to 'Science Achievement' in the Philippines. This weight is systematically positive, while the overall value is negative and fits the other countries' values well. The other high coefficient is for 'Literacy of Home' in Korea. The overall value is .18 while the Korean value is .41. This means that home literacy has a much larger effect on achievement than is true in other countries. This may be due to the fact that home literacy was measured in Korea only with the Word Knowledge test. The remaining three large discrepancies are negative. 'SES' in Canada (Fr.), 'Literacy of Home' in Hong Kong, and 'Experiments' in Korea, have essentially no effects on

TABLE 8.1 The effects of constructs on science achievement (Population 2)

Country	SES of Home	Literacy of Home	Sex of Student	Like School	Teaching Style	Experiments	Classroom Effort	Homework	School Science Equipment	Views of and Interest in Science	R^2
Australia	+++	+++	--		--	+		+		+	.35
China	++	++	---		-	+	+	++	++	+	.23
England	+++	+++	--		--	+		++	+	++	.37
Finland	++	++	--	+	-	+		--		+++	.19
Hong Kong	+	+	--		--	+++		++	++	++	.28
Hungary	+++	+++	--	+	--					++	.23
Italy (Grade 8)	+++	+++	--		-	-	+	++	+	+	.31
Italy (Grade 9)	+++	+++	--		+	-	++	++	++	--	.16
Japan	na	+++	--	++	--	+	+++	+		+	.27
Korea	++	+++	--	+	-	-	+		+++	+++	.36
Netherlands	+	++	---	+	--		++	+	++		.37
Nigeria	++	++	--	+	-	-		++		+++	.15
Norway	na	+++	---	na	--	+++	na		+	na	.16
Philippines	+++	++	--	++	--	+	+	+	+	+++	.27
Poland	++	++	--		-			+		++	.12
Singapore	+++	+++	--		-	+		++		+++	.40
Sweden (Grade 7)	++	+++	--	+	-	+++	++			+	.23
Sweden (Grade 8)	++	++	--	+	-	+	++	-		+++	.27
Thailand	+	+	--		---	++	+	++	-	++	.17
Signs for 19 countries	17+	19+	19-	9+	18- 1+	12+ 3-	10+	13+ 2-	8+ 1-	15+ 1-	

achievement, whereas these constructs have positive effects in all other countries. The important finding here is not, however, the discrepancies between countries, but the fact that countries are extremely similar in the structural relations between achievement and its determinants. Out of the 76 distinct coefficients relating constructs directly to achievement, 71 of those values could be treated as constant over all countries.

Following the analyses of the seven countries given above an analysis was undertaken on 19 data sets covering 17 countries (Italy had two populations: Grade 8 and Grade 9; Sweden had two populations: Grade 7 and Grade 8). Table 8.1 presents the results of the influences of ten constructs on science achievement. Using the conventions of a single + (plus) or - (minus) sign representing a weak effect (+: .05 to .09; -: -.05 to -.09), ++ or -- representing an effect (++: .10 to .19; --: -.10 to -.19), and +++ or --- representing a strong effect (+++: .20 or higher; ---: -.20 or lower) the table was constructed to indicate the effects on student science achievement within each country. The final column represents the number of times there was an effect. The convention 'na' means that no data were available and no entry means that the coefficient lay between ±.04, indicating that there was no effect.

The constructs are the same as for the analysis presented in Figure 8.1 with the following exceptions.

Influences on Student Achievement

Socio-Economic Status of Home	'Mother's Education' was dropped.
Like School	'Enjoy everything at school' and 'School most enjoyable' dropped.
Teaching Style	Only the variables 'Students choose topics for study', 'Teacher uses student ideas' and 'Students own problems and solutions' were used.
Time	was not used
View of and Interest in Science	Only five variables were used. These were: 'Science relevant for everyday' 'Science important for country development' 'Worth spending money in science' 'Science is for creative people' 'Science helps you learn more about the world'.

Factors Influencing Achievement in Population 2

The Home   In all countries the socio-economic status of the home and the literacy of the home had a very strong influence on students' science achievement. Indeed, the home is still the most powerful influence on achievement. This reinforces the old adage that the best thing a child can do is to choose good parents. It must be remembered that there is more variance between homes than between schools. After all, there is no entry requirement to become a parent. Teachers must pass examinations to become a teacher and schools falling below a particular standard are closed. If there was less variance among homes the effect would be weaker. It is to be assumed that the parents who are richer and with more education have more books at home, provide a richer vocabulary context and encourage their children in their science learning more than do poor homes with few books.

Sex of Student   This path is a relationship rather than an influence but in all cases boys performed better than girls. However, it is clear that how girls perceive schooling, are treated in school and so on does differ from boys in many schools. Much work remains to be done to unravel this problem.

View of and Interest in Science   Those who viewed science as being important and useful achieved better than those who did not do this. From another analysis undertaken by Morgenstern (1990 op cit.) the two constructs most influencing a positive view of science were 'Science Achievement' and 'Like School'. Thus, there is an interaction between 'View of Science' and 'Achievement' and they reinforce each other.

Experiments   In most countries this construct influenced achievement. This was particularly the case in Hong Kong and Norway and it is to be presumed that there is large variance in those countries between schools in the amount of practical work they can or do undertake.

Classroom Effort   In some ways this construct is a proxy for motivation and in many countries it has a direct influence on science achievement. The same was true for *homework*.

Teaching Style This construct is negatively associated with science achievement in all countries except for Italy (Grade 9). The data about 'Teaching Style' were collected from students and not teachers. The data constitute students' perceptions of what is happening in the classroom. However, it may be the case that the more the students perceived a lack of teacher direction the worse they achieved. This is clearly a construct that warrants further investigation in each country.

Like School In general, girls liked school more than boys but, whichever the sex, liking school had a real and positive effect on science achievement in many countries. The effect was moderate in Japan and in the Philippines.

School Science Equipment In eight countries those students in schools with more science facilities had a positive effect on science achievement. In general, small schools had fewer facilities than larger schools.

There are two overall points to be noted. The first is that in most cases, the influence that any one variable or construct has on science achievement is associated with variance in that variable or construct. It points to features of schooling in science where there are differences amongst students or schools. Take Hong Kong as an example. The largest influence is 'Doing experiments'. However, 'School Equipment' also has an influence and the two are coupled. If the Hong Kong authorities ensure that those schools with no facilities or poor facilities receive more facilities and if the teachers in those schools have the students do more experiments, then the influence of these features of schooling on science achievement will decrease. In the next study it is probable that other features where there is still variance will become more important. It is in this way that planners can learn about the weak links in their system and then correct them. It will also be seen from Appendix I that both *homework* and *instructional time* had a considerable effect on achievement in some countries and a very small, but positive effect, in other countries. This again is primarily due to the amount of variation in the two sets of countries.

The second is that the amount of variance explained (R^2) is relatively low ranging from 12 percent in Poland to 37 percent in England and the Netherlands. This means that the constructs in the model which have paths to science achievement are not 'explaining' even half of the variance among students. There are, in other words, other features of schooling and of student attributes which are not in the model which must explain more of the variance. It can be that other researchers using more of the variables in the data set will be able to 'explain' more of the variance in secondary analyses.

Reference

1. Keeves, J. P. 1991 *Changes in Science Education and Achievement 1970-1984*. Pergamon, Oxford.

2. Morgenstern, C. 1990 *Determinanten naturwissenschaftlicher Leistung in allgemeinbildenden Schulen im internationalen Vergleich* (Determinants of science achievement in general schools in an international perspective). Unpublished Diplom thesis. Faculty of Sociology, University of Hamburg.

9
Summary of Results

This chapter is divided into two parts. Part 1 reviews each system of education in terms of the data supplied. Part 2 summarizes the major findings.

Part 1: Comments on Each System of Education

Australia

The lower population in Australia was drawn from Grades 4, 5, or 6 since the 10-year-olds were to be found in different grades according to the structure of education in the different states in Australia. The average age was 10 years six months. Forty-four percent of their teachers were male and the rest female. The teachers had an average age of 33 years and had about 3.5 years of post-secondary education. The teachers had a great deal of autonomy in what they taught and for the most part they were general class teachers. About five percent of the school's timetable was devoted to the teaching of science. The children were in class sizes of 25 to 30 students.

At Population 2 level the class sizes remained the same but the percent of instructional time devoted to science rose to 15 percent (about 3.8 hours per week). Seventy-four percent of the science teachers were male and were specialist science teachers with 4.5 years of post-secondary education. Science was taught as general science.

At Grade 11 science is commonly taught as the separate subjects of biology, chemistry, and physics. By the terminal grade in school only 39 percent of an age group was in full time academic education. About one half of these students studied biology and a third studied chemistry and/or physics. The students studied five subjects each on average. Seven and a half percent of the biology students used a language at home that was not English and 16 percent of the chemistry and physics students did not speak English at home. The average age of twelfth grade students was 17 years three months which was about one year younger than in most other systems.

For biology about half the teachers were men and half women whereas for chemistry and physics nearly all of the teachers were men. These teachers had had nearly five years of post-secondary education and the average attendance at in-service teacher training courses was two days per year. The teachers taught about 18 hours per week and spent nearly the same amount of time in preparing lessons and marking homework. The schools were well equipped with laboratories and nearly all science lessons were conducted in laboratories.

The science achievement of the 10-year-olds was good relative to most other countries but their achievement at Populations 2 and 3 was only average compared with other countries and there is clearly room for improvement. At Population 1 level, their chemistry scores were somewhat low; at Population 2 level their earth science scores were slightly above average; and, at Population 3 level their biology scores were relatively low compared with their other scores.

However, a higher proportion of the age group studied biology than physics and/or chemistry.

Canada (English)

The youngest population in Canada (English speaking) was in Grade 5. Their average age was just over 11 years. Nothing can be said about any of the teachers at any level in Canada (Eng.) because no data were forwarded for analysis. The Grade 5 students were in relatively small schools (an average size of 274 students). All schools were coeducational and reasonably well catered for in term of teachers and science equipment. The curriculum was determined by provincial and school boards but the school principal and teachers decided on the textbooks which were to be used. Only 60 percent of the classes were taught by the class teacher whereas the rest were taught by other teachers than the class teachers. A small percentage of schools had specially trained science teachers.

At the middle population level (Grade 9) students had an average age of 15 years. No data were available on the weekly hours of instruction in science, but the students reported doing an average of two and a half hours science homework per week. Eighty percent of the science teachers were male and the schools were well equipped with laboratories.

At the top end of the school system 68 percent of an age group was enrolled in Grade 12. Twenty eight percent of an age group studied biology, 25 percent chemistry and 18 percent physics. On average, they studied six subjects. The average age of the students was 18 years and three months. The average school size was about 800 students. Ninety seven percent of the students were in mixed schools and 85 percent of the science teachers were male. The schools were well-equipped with laboratories (an average of 4.7 laboratories per school). They were used three quarters of the time with the assistance of about 1.5 laboratory assistants.

The achievement scores of Canada (Eng.) at Population 1 were relatively high (compared with other countries) although their chemistry score was slightly lower than their scores in the other branches of science. At Population 2 level, their scores were average and at Population 3 level their scores were relatively low for biology and physics and average for chemistry. In general, however, their relative achievement declined from one population level to the next.

Canada (French)

Population 1 students were in Grade 5 and had an average age of just over 11 years. The average class size was 25 students. The teachers were mostly women with an average age of 41 years. They had about four years of postsecondary education but only seven percent of this had been devoted to science education. They received an average of one day per year inservice training. They taught 21 hours per week but less than two hours was devoted to teaching science. Relatively little practical work in science was undertaken. The pupil-teacher ratio was 21.0 (as compared with 18.5 in English-speaking Canada). Eighty-two percent of those teachers teaching science were general class teachers.

Summary of Results

At Population 2 the profile of the French-speaking Canadian students was similar to that of the English-speaking Canadians. Science subject specialization began at Grade 9 where biology and physics were sometimes taught simultaneously. They were in classes with an average size of 30 students. The amount of time spent on science homework was less than that of their English Canadian counterparts. Three quarters of the science teachers were men with an average age of 40. They had 4.8 years of postsecondary education of which 1.3 years were spent studying science. They had an average of 16 years teaching experience and received three days a year inservice training. Eighty-seven percent of their 19 hours weekly teaching load was spent teaching science. The average school size of 939 was 1.5 times higher than that for English-speaking Canadian students. Seventy-seven percent of science classes were conducted in the laboratory.

At Population 3 level the students in Québec were from Grade 11 whereas the students in the few francophone schools tested in other provinces were from Grade 12. Sixty-seven percent of an age group were in academic streams. If the vocational tracks were included the percentage would be 79 percent. The average age of the students at the time of testing was 17 years two month. The average number of subjects studied was six. Three quarters of those studying biology were girls whereas for physics and chemistry, the proportions were about 50 percent. Depending on the subject studied nine to 11 percent used a language at home different from that used in school. For biology and chemistry, four fifths of the teachers were male and for physics nearly all were male. The teachers had an average age of about 40 years. They had nearly five years of post secondary education and received between three and four days of inservice training per year. They had the highest total teaching load per week of all countries (about 21 to 23 hours) and spent the least amount of time preparing lessons and marking students' work (9 to 10 hours). Only about 15 percent of teachers perceived their science teaching to be hampered severely by lack of equipment. Most schools were mixed and had, on average, more than 1,000 students per school. Two thirds of all teachers were male and 85 percent of all science teachers were male. The schools have an average of 5.3 laboratories per school and 1.1 laboratory technicians, a little less than Canada (Eng.).

The achievement of Population 1 students was relatively high in all areas of science.

At Population 3 level (which it must be remembered was Grade 11) the achievement levels were all relatively low. This is a further example of relative performance decreasing as one ascends the school system.

China

China tested only at Population 2 level. They tested Grade 9 where the average age was high at 15 years and eight months. Only 37 percent of an age group was in school. China tested in 1985 and only in the three provinces next to the capital which is probably a more privileged section of their school population than those at greater distances from Beijing. Science specialization began in Grade 7 where separate science subjects were taught simultaneously. The average number of children per family was two. From other information, it would appear that it is the urban families which tend to have one child and the rural families two and three. The families had, on average, about 80 books. The students did 7.6 hours a week of homework of which two thirds was on science. There were about equal numbers of men and women science teachers with the highest average age (47) amongst all countries. They also had the lowest (1.5 years) number of years of postsecondary education. All schools were co-educational and had on average 805 students. The number of laboratory technicians was the highest (average of 3.3) at Population 2 level while the number of laboratories per school was 3.9. About 50 percent of the time was devoted to the conduct of science experiments.

The achievement of the students in the three provinces was relatively high.

England

The Population 1 English students were from Grade 5 and had an average age of 10 years three months. In general, their homes had almost as many books (212) as the Australian homes (235). They were in classes of about 28 students. Sixty percent of their teachers were women with an average age of 39 years. The teachers had 3.5 years of postsecondary education and one day per year inservice training. On average they had 13.4 years of teaching experience. They devoted 38 percent of their science teaching time to practical work. Science was taught frequently by the class teachers (68%) to their own classes. Only about 10 percent of primary science was taught by teachers who had specialized in science. The remaining 22 percent taught science to two or more classes, other than their own. The schools were relatively small (an average of 245 students) and the pupil-teacher ratio was 25. In 1984 the teachers and school principal determined the school science curriculum while the class teachers made the choice on use of texts and equipment. This last point is important because while one hears much about the British primary school science program, it was clearly up to the teachers to teach what they wished, but more recently, new policies have been introduced across the country.

The Population 2 students were in Grade 9 and came from families with an average of 2.8 children and 'literate' homes having about 200 books per family. They were in classes of about 28 students receiving six hours of science instruction per week and they did somewhat less science homework than students in most countries with the exception of Sweden (about one hour only per week). Most of the science teachers were men with an average age of 36 and 12 years teaching experience. The teachers had had 4.3 years of postsecondary education and received on average about two days per year inservice training. They were in schools with an average size of just over 900 students. However, eight percent of schools were boys' schools, seven percent girls' schools and the rest coeducational. The schools were well equipped with laboratories. It is to be noted that the content syllabus of the curriculum was completely in the hands of the teachers but presumably controlled to some extent by the content of the external examinations the students were meant to take. Science subject specialization generally began at Grade 9. England claims to have seven laboratories or classrooms equipped for science teaching per school, manned by an average of 2.1 laboratory technicians. About two thirds of the time was devoted to doing experiments.

Summary of Results

The students in Grade 13 were 18 years old and studied, on average, only three main subjects. About 20 percent of an age group was enrolled in school (colleges of further education were excluded) at this level and of these only one quarter studied science. Sixty percent of those studying biology were girls in contrast to 67 and 78 percent of boys studying chemistry or physics. Nearly all expected to have three years of university education. The average class sizes were small (10 to 12 students per class). The students spent eight to 10 hours a week studying science and a high proportion of their homework was spent on science (8.4 out of 9.1 hours of all homework). Thus, England has a highly select group of students specializing to a great extent in science. For biology, two thirds of the teachers were men and for chemistry and physics the figure was over 80 percent. All have had four to five years of postsecondary education and received two to three days of inservice training per year. The teachers gave about one third of their teaching time to practical work and nearly all lessons were taught in laboratories. The average number of laboratories per school (8.8) was the highest of all countries. There were also 2.8 laboratory technicians per school. Only 73 percent of the schools were mixed schools; 18 percent were boys only and nine percent girls only. The average size of school was 900. Half of the general teaching staff was female but 70 percent of the science teaching staff was male. This was in the context of about 16 hours per week of science instruction. Not only is the English (about 86 percent) group an élite but it would also appear to spend a lot of time studying science in schools with good facilities.

England shows the reverse picture to those of the two Canadas. The performance was relatively low at Populations 1 and 2 but high at Population 3. At Population 1 level, the scores in chemistry and physics were relatively lower than those for biology and earth sciences. At Population 2 level, scores were relatively low in all branches of science. At Population 3 all scores were high.

Finland

The Population 1 students were from Grade 4 and had an average age of 10 years and 10 months. They came from homes with an average of 2.5 children per family. Relative to Norway and Sweden (224 and 207 books), they had few books (142) at home. However, it should be noted that public library services exist everywhere in the country and are intensively used. They began school at the age of seven and had relatively few hours of total instruction per year. Sixty percent of the teachers teaching science were men. The teachers teaching science had four years of postsecondary education of which 14 percent of the time was devoted to science. They had half a day of inservice training per year. They taught only 18 hours per week of which 3.2 hours was devoted to teaching science. The schools were small (average of 195 students) and the pupil-teacher ratio was small (about 18.5). Sixty percent of schools had science taught by the class teacher but 40 percent had several "specialist" teachers teaching science. The range of science subjects and the content of science is controlled nationally but the choice of texts is a school board and teacher matter. It is important to note that it is in Grade 4 that science is divided into two subjects. This is the earliest separation point of all countries in the study.

At Population 2 level the students received three hours science per week. The average class size of 16 students was the smallest of all countries. So was the average school size of about 500. The students did relatively less total homework (5.7 hours) per week when compared with most other countries. Likewise, the science homework (two hours per week) was only one third of the total homework time. The percentage of students saying they did experiments in science was very low (21%). This is second lowest only to Poland (11%). There was about an equal number of men and women science teachers with an average age of 39 years and 12.5 years of teaching experience. They, like the French-speaking Canadian teachers, had an average of 5.7 years of postsecondary education and two thirds of that time was spent studying science. They received an average of three days a year inservice training. They had a low teaching load of 17 hours per week of which 71 percent was spent teaching science. They spent 14 hours per week preparing lessons and marking students' work. The total hours of instruction per year was the lowest (874 hours) of all countries. Their laboratory provision was above average. No schools had any laboratory technicians.

Forty-one percent of an age group was enrolled full-time in academic tracks at Grade 12 level. All of them studied biology but only 16 percent of an age group studied chemistry and 14 percent physics. All studied some eleven subjects. Of those students studying biology or chemistry about 53 percent were girls. In contrast, for physics, there were only 37 percent girls. The average class sizes were 26 for biology and 22 for chemistry and physics. The percentage of students reporting that they did experiments was extremely low compared with other countries (12 to 24 percent as compared with 80 to 90 percent in other countries). Again, the number of hours spent studying science (four to five hours) was very low compared with other countries. Unfortunately, neither of the other two Nordic countries supplied information on this variable. Hence, it was not possible to see whether this was a Nordic phenomenon. Yet again, the proportion of all homework time devoted to science was consistently lower than other countries across all the three science subjects. About half of the teachers teaching biology and chemistry were female but for physics 75 percent were male. The biology teachers were, on average, 45 years old. This was higher than for other countries, and the biology teachers were older than the chemistry and physics teachers. All teachers had had nearly six years of postsecondary education of which they reported two thirds was spent studying science. They also received an average of 3.5 days inservice training per year. The average teaching load was 16 hours per week. The time spent preparing lessons and marking homework was lower for biology and chemistry than that for physics. Only 17 percent of biology teaching time involved practical work but this was 50 percent for chemistry and 31 percent for physics. Only about 10 percent of the science teachers perceived the lack of equipment as a severe handicap to their teaching. The average school size (535) was small compared with most other countries. All schools were mixed. Finland's picture was somewhat similar to the two Canadas. Population 1 scores were amongst the highest. The same *tended* to be true for Population 2 but at Population 3 all scores were relatively low.

Ghana

Ghana did not test at Population 1 level. At Population 2 the Ghanaian students came from the academic secondary schools which enrol only six percent

Summary of Results

of an age group. In Grade 9 the average age was about 16 years. They were, in a sense, the élite of the system. Seventy percent of the students were boys. The students were from large families having an average of 130 books in each home. The class size was about 40. The allocated instructional time for science was about six and a half hours a week. The students reported doing eight hours of homework for all subjects but claim that six of the hours were spent on science homework. About 70 percent of the science teachers were men with an average age of 30 years and six years teaching experience. They received 4.2 years of postsecondary education of which three quarters was spent studying science. They also received one day of inservice training per year. They taught an average of 17.2 hours per week of which 86 percent was spent teaching science. Twelve percent of the schools were boys' schools, 10 percent girls' schools, and 78 percent mixed schools. The school size was relatively small (average of 579). There were, on average, 2.4 laboratories per school and 2.5 laboratory assistants. The percentage of time that school laboratories were used was only 54. About half the time in science instruction involved doing experiments.

Only 1.2 percent of an age group was enrolled in Grade 13 in Ghana. Only 0.2 percent of an age group studied biology, and 0.6 percent chemistry and physics. The students were 18 years 10 months old and studied only three subjects. More than 80 percent of the students were boys. They expected to continue to university for three to four years of study. The class size was about 30 for chemistry and physics. All did experiments and reported studying 20 hours of science per week. Nearly three quarters of their homework was science homework. The data on teachers were not reported in this volume since so few teachers answered the teacher questionnaires that it would be folly to infer anything from them. Ghana did not collect school data.

Population 2 students scored amongst the lowest group of countries in all branches of science. The Population 1 scores were slightly lower than Japan and Korea's. At Population 3, the scores were relatively high (biology and chemistry) and average (physics).

Hong Kong

Hong Kong is a large metropolis. There were 3.7 children per family and the homes had relatively few books.

The Population 1 students were in Grade 4 and had an average age of 10 years and five months. The average class size was 38 students. The teachers had an average age of 37 years and they had two years of postsecondary education which included little science. They had an average of 15 years teaching experience. A few of them received quite a lot of inservice training but many had had none. They taught an average of 18 hours per week but just two periods of 30 minutes each on science. Little time was given to practical work. The pupil-teacher ratio was high at 30.2.

At Population 2 level, the Grade 8 students received 2.7 hours per week instruction in science in classes of 40 students. In Grade 9, about half of the schools opted for separate sciences rather than integrated science. The other half began specialization at Grade 10. The students did six hours of homework per week of which about half was for science. Fifty-four percent of the science teachers were female even though 66 percent of all teachers were men. The teachers had an average age of 31 years and 7.7 years of teaching experience. They had 2.8 years of postsecondary education of which 80 percent was spent studying science. They had a teaching load of 18.6 hours per week and 65 percent of this time was spent teaching science. The schools were large with an average of 1,175 students. Eleven percent of them were boys' schools, eight percent girls' and the rest were mixed. The schools were well equipped with laboratories, manned by an adequate number (2.5) of laboratory assistants. Two thirds of the science time involved the conduct of experiments.

At Population 3 level, data for both Form 6 (Grade 12) and Form 7 (Grade 13) were presented. Form 6 is the terminal Grade for students proceeding to the Chinese University and Form 7 is the terminal Grade for those going to the University of Hong Kong where the medium of instruction is in English. Twenty seven percent of an age group was enrolled at Form 6 and 20 percent at Form 7. At Form 6 level, 12 percent of an age group studied biology and 20 percent chemistry and 20 percent physics. At Form 7 level only seven percent studied biology but 12 percent studied chemistry and physics. The average age of Form 6 students was slightly over 18 years of age and Form 7 slightly over 19 years of age. The average number of subjects studied in Form 6 was six and in Form 7 it was five. Over two thirds of the science students were boys. At Form 6 level, chemistry, and physics students studied these subjects for 11 hours per week and at Form 7 level this was 13 hours per week. However, Form 6 biology students studied this subject for 13 hours a week and Form 7 biology students for 16 hours of science per week. Most of their homework was in the sciences. Nearly 100 percent reported doing experiments. All science course teachers in both forms were graduates of the University of Hong Kong. Sixty percent of the biology teachers were men; for chemistry 80 percent were men, and for physics 94 percent were men. The average age (31 years) of teachers was lower than in other countries. Most had four years of postsecondary education of which over 80 percent was spent studying science. The average teaching load was between 17 and 18 hours per week and all indicated that they spent more time (19 to 20 hours) than this preparing and marking. About one third of teaching in all subjects involved practical work and over half of the teaching time was undertaken in laboratories. About twelve percent of biology and chemistry teachers felt that they were hampered by lack of equipment in their teaching, but for physics this figure was somewhat higher. The average size of school was around 1,000 students. About 80 percent of schools were mixed and of the sex-segregated schools there were about twice as many boys' as girls' schools. The total teaching staff was equally split between male and female but 71 percent of science teachers were male. The schools had an average of 4.3 laboratories per school and nearly three laboratory technicians per school. The overall picture was one of a studious élite of students who were paced hard by the teachers in relatively good conditions for science learning.

At Population 1 level, the students were average in biology but relatively poor in the other subject areas. At Population 2, they had average scores in earth science and were relatively poor in the other areas. However, at Population 3 level in both Forms 6 and 7 and with 12 to 20 percent of an age group they performed outstandingly. Indeed, their Population 3 scores are a good example of what is possible. However, little attention would appear to be given to science education up to the end of Form 2 in Hong Kong and one wonders about the apparent emphasis on the science education of the élite at the expense of mass education lower down the system.

Hungary

Population 1 students were in Grade 4, had an average age of 10 years and three months, were from literate homes with an average of 2.4 children per family. The teachers had three years of postsecondary education of which 46 percent had been the study of science. The teacher load was low (16.6 hours per week). Of this, about nine percent of the time was allotted to the teaching of science. All teachers had an average of 3.8 days per year inservice training which is the highest of all countries. The pupil-teacher ratio was the smallest of all countries (about 11.8). But they had the highest percentage use (77%) of laboratory facilities. The curriculum was controlled nationally but it was the teachers who chose the textbooks and equipment.

Population 2 students were from Grade 8. Only 92 percent of an age group was enrolled in school. The average number of children per family was 2.3. In this study only China had a smaller number of children per family. On average, the students received 4.2 hours of science instruction per week. They reported doing nine hours of homework per week of which six were spent on science. Science subject specialization began at Grade 6. It replaced five years of general science (knowledge of surroundings). Biology and physics were taught in Grades 6 to 8 while chemistry was taught in Grades 7 and 8. The students were in class sizes of about 28. Only 22 percent reported that they did experiments. Seventy percent of their science teachers were female although 80 percent of all teachers (science and others) were female. The teachers had an average age of 38 years and nearly 16 years of teaching experience. They had had four years postsecondary education of which 2.6 years were spent on science. They received an average of 4.5 days per year inservice training (again, the highest average of all countries). Their overall teaching load was low (16.4 hours per week). Of this, 62 percent of their time was spent teaching science. They prepared and marked for 18.4 hours per week. The percent of time spent on practical work was about one quarter of all instructional time in science. All schools were coeducational and the average size of school was 580. There were 2.7 laboratories per school which were used 78 percent of the time. As stated before, they spent only one fifth of their science time doing experiments which is very low compared with all other countries. There was one laboratory assistant per school.

Although 40 percent of an age group was enrolled in Grade 12 in Hungary only 18 percent were in academic tracks. All of these 'academic' students studied general science but some opted to do 'enrichment courses' in biology, chemistry or physics. It is these students who formed the target population of this survey. Three percent of an age group took these 'enrichment courses' in biology, one percent in chemistry and four percent in physics. They were 18 years old and studied at least nine subjects. Sixty-four percent of the students in biology and chemistry were girls but only 29 percent of those studying physics were girls. The biology and physics classes had 30 students in them but the chemistry classes only 11 students. Again, the biology and physics students studied only 12 to 13 hours a week science, but the chemistry students had 16.5 hours per week. About less than half of their total homework time was devoted to science. As already stated

64 percent of the biology students were female. Almost the same percentage of biology teachers were female; 50 percent of the chemistry teachers and 38 percent of the physics teachers were female. The average teacher was in her early 40s, had had 5.5 years of postsecondary education of which close to two thirds to four fifths was devoted to biology or physics but only two thirds to chemistry. The teaching load for biology teachers was 15 hours a week, for chemistry, 16 hours and for physics, 17 hours. All reported spending slightly over 20 hours per week preparing lessons and marking students' work. The percentage of time teachers reported they spent on practical work was low, ranging from 24 percent for biology to 21 percent for physics and chemistry. Twenty-four percent of biology and chemistry teachers and 16 percent of physics teachers complained of lack of equipment. The average size of school (all were mixed) was small with 388 students per school.

At Population 1 level the students were average in achievement in physics but high in other science areas. At Population 2 the students scored the highest of all countries. At Population 3 they were high in biology and physics but only average in chemistry, despite the fact that the Population 3 chemistry students had more hours of science instruction than biology and physics students had.

The Hungarian performance was good and especially for mass education at Population 2 level. And the performance of the bottom 20 percent and the performance of the lowest school was good. However, it should be noted that only 92 percent of the age group were in school at the Grade 8 level, and the evidence available indicates that some less able students had left school by this stage.

Israel

The Population 1 students were from Grade 5 and had an average age of about 11 years. They came from homes with relatively few books (about 14), the fewest reported in this study. The schools offer 11 years of compulsory education and at Grade 5 had about 200 days per year schooling. The students received two hours a week instruction in science. They had mostly women teachers (93%) with 2.6 years of postsecondary education of which nearly half the time was spent studying science. The teachers also had three days per year inservice training. Their teaching load was 23 hours per week of which nearly 10 hours was spent teaching science. In other words, some of the teachers were 'specialist' science teachers. Forty two percent of the science teaching time was spent on practical work and 55 percent of the time in laboratories or special purpose rooms. There was a national curriculum devised by the national curriculum centre in conjunction with teachers.

Israel offered two science subjects at Junior High, namely 'Life science' and 'Physical science'. There were slightly more girls than boys enrolled in Grade 9. They were from families with an average of 3.8 children. They received 6.3 hours of science per week and spent about 8.4 hours on all homework of which 3.9 hours were spent on science homework. Fifty percent reported doing experiments in the laboratory. Thirty-one percent of science time was practical work. Schools had an average of 3.5 laboratories and 2.4 laboratory technicians. The teachers had an average age of 35 years and 11 years teaching experience. They had 4.8 years postsecondary education and three quarters of that time was spent on science. They had nearly five days per year inservice training. Their total teaching load was 23 hours per week and they spent nearly four fifths of that time teaching science. Nearly forty percent of the science teachers were male.

Population 3 students were in Grade 12. At that level 65 percent of an age group was enrolled in school. Twenty percent of an age group studied biology, eight percent chemistry and 12 percent physics. The average age was 17 years and seven months and the students study an average of seven subjects. All of the science students expected to go to the university for another three years. They studied three hours of biology, and an average of about 4.6 hours of chemistry or physics per week. Between half to two thirds of the total homework time was spent on science. No teacher data were available for Israel. From the school data that were available, it would appear that 63 percent of all teachers were female and that 58 percent of all science teachers were female. Israel provided a relatively sizeable amount of time (14.7 hours per week) for science instruction. There was an average of 3.7 laboratories per school which were in use 44 percent of the time. There were 2.6 laboratory technicians per school.

The performance of Grade 5 students was low compared with other countries. At Population 2 level achievement in chemistry and earth science was high, average for biology and low for physics. At Population 3 level Israel was for all science subjects in the middle of the range of the systems considered.

Italy

Italy's Population 1 were all students in Grade 5. Their average age was 10 years nine months. Nearly all of the teachers were women but with an average age of 45 years and 21 years of teaching experience. Their postsecondary education was the lowest of all countries (one year). Their total teaching load was 20 hours per week and an average of 3.5 hours was devoted to the teaching of science. They did very little practical work. The schools had a low pupil-teacher ratio (13.7).

At Population 2, Italy tested two grades: Grade 8 and Grade 9. Grade 8 is the last year of the *scuola media* and compulsory schooling. Twenty-eight percent of the age group then dropped out of school and the other 72 percent entered various types of schools. Slightly more boys than girls left school. Both were from larger and poorer families. The Grade 8 students had an average age of 13 years 11 months and the Grade 9 students 14 years and eight months. Grade 9 students were in larger class sizes than Grade 8 (27 to 23) but received more science instruction (4 hours compared with 3.6 hours per week) and did more homework. In both grades there was an equal number of men and women teachers. Both sets of teachers had four years of postsecondary education of which two thirds of the time had involved science. Grade 8 schools had an average of less than one laboratory per school but Grade 9 schools had three laboratories per school. Nevertheless, the time spent on practical work and in laboratories was very low. The teachers appeared to have great freedom in deciding the content of what they taught and the textbook they used.

In the terminal grade, only 34 percent of an age group was enrolled in school. Four percent of an age group studied biology, one percent chemistry and 13 percent physics. Their average age was 19 years and they each studied seven subjects. Many more boys than girls studied science at this level. The hours of instruction in science they received in any one subject was about 6 hours per week. The science teachers' (61 percent female) average age was 40 years and they had four years of postsecondary education with at least two thirds of that time devoted to science. The average number of days inservice training was high (three to four days per year). The average total teaching load was 17 to 18 hours per week but the time spent preparing lessons and marking students' work was low (10 to 12 hours per week) compared with other countries. Again, the time spent on practical work was low.

At Population 1, Italy was in the middle of the achievement range, was relatively good in biology, chemistry and earth sciences but poor in physics. In Grade 8 the students were generally low in all subject areas but in Grade 9 the students were relatively good in biology and earth sciences, average in chemistry and poor in physics. At Grade 13 the students performed poorly compared with other countries in all branches of science. From an average performance at Population 1 level, the performance declines in the higher levels of schooling. Indeed, there would appear to be problems in science education in Italy.

Japan

Population 1 students were in Grade 5. The system provided 210 days of schooling per year. The average age of the students was 10 years and seven months. The average class size was 38 which was close to Hong Kong and second only to Korea (55). Half of the teachers were men. The teachers' postsecondary education lasted an average of 3.3 years of which 18 percent was devoted to science. They received 1.7 days inservice training per year. The teaching load was 26.8 hours per week which was the highest of all countries. Of this, they reported spending 3.5 hours per week teaching science although the official instructional time for science was three hours per week. Since 16 percent of the schools used specialist science teachers who taught more than one grade or class, this accounts for the extra 0.5 hours. Just over 40 percent of the science teaching time was spent on practical work and in laboratories. It was reported that every school has at least one science room or laboratory. This high combination of practical and laboratory work existed only for Japan and Korea. The average school size was 757 students and the pupil-teacher ratio of 29.8 was close to Hong Kong and Nigeria but much smaller than Korea. The schools used their laboratories 58 percent of the time. The curriculum was a national one but the choice of textbooks was up to the school boards and the choice of equipment primarily a matter for the principal and teachers.

Grade 9 students in Japan were in relatively large classes (42 students per class). This is close to Hong Kong (40), and Thailand (42) but still far lower than the Philippines (48) or Korea (65). They studied science for four hours a week and did science homework for an average of 2.3 hours per week. This was about 27 percent of all homework. Eighty-six percent of the science teachers were men even though men constitute only 67 percent of all teachers. The teachers had an average age of 38 years and nearly 16 years teaching experience. They had nearly four years of postsecondary education of which 60 percent of that time was spent studying science. They received four days of inservice training per year. The total teaching load was 18.5 hours per week and nearly all of the this time was spent teaching science. This implies that most science teachers were specialized. All schools were coeducational and the average school size was 784. There were two science laboratories per school and they were used 68 percent of the time. About 50 percent of time was spent doing experiments.

In Grade 12, 89 percent of an age group was enrolled in school. However, 63 percent of an age group was enrolled in academic schools and it is this group which constituted Population 3. Twelve percent of an age group studied biology, 16 percent chemistry, and 11 percent physics. The average number of subjects studied was seven and the average age of the students was just over 18 years. The biology students studied science subjects for 4.8 hours per week, the chemistry students 6.4 hours, and the physics students 6.8 hours. All had about 10 to 15 hours of total homework per week of which they reported a quarter (biology) to about 35 percent (chemistry and physics) were spent on doing science homework. The teachers were 40 to 43 years old on average and nearly all were male. All had just over four years of postsecondary education of which nearly two thirds was spent studying science. (At the university level students tend to study general subjects for the first two years and then specialize.) The teachers received three to four days of inservice training per year. The average teaching load was 16 hours per week but 19 to 21 hours were spent on preparation and marking. Compared with other countries, the teachers reported spending a relatively small percentage of their teaching time (15 to 22 percent) on practical work but with some 40 to 50 percent of their lessons being held in laboratories. It is typical for science lessons to be held in laboratories but without practical work being done by students. Teachers may demonstrate an experiment for 10 minutes in a one or two hour science lesson. Twenty percent of chemistry teachers perceived their teaching to be hampered by lack of equipment but for biology and physics teachers this was about 38 percent. Seventy-eight percent of the schools were mixed, 17 percent girls only, and five percent boys only. The average size of school was 1074 students. Eighty percent of all teachers were male and 95 percent of all science teachers were male. There was an average of 4.4 laboratories per school which were used 53 percent of the time with the assistance of 1.4 laboratory technicians. The striking facts are the relatively few hours devoted to science, the male dominance of the teaching force at the high school level, and the relatively higher than expected proportion of teachers who perceive their schools to lack the facilities necessary to teach science.

Population 1 and 2 students achieved extremely well compared with other countries. The differences among schools was also very low. But at Population 3 level the picture was differentiated. The Japanese students scored surprisingly poorly in biology, in the middle of the range in chemistry and well in physics. The differences in achievement between schools were also high. Many of the twelfth graders studying science did not see themselves entering a career involving science. However, across all countries the highest proportion of the age group planned to continue with the study of science at the postsecondary school stage in Japan.

Korea

In 1984, Korea invested 20.1 percent of its state budget in education. This was the highest of all countries. It also had 220 days of schooling per year at all grade levels.

The Population 1 students were from Grade 5 and in classes of about 55. The average age was 11 years two months. At Grade 5, 58 percent of the teachers were male. They were, on average, 35 years old with 13 years of teaching experience. They had had two years of postsecondary education of which 13 percent was devoted to science.

The inservice training in science was 2.3 days per year. Their teaching load was 21.2 hours per week of which three hours were allotted to teaching science. Thirty-eight percent of their science teaching time was given to practical work in laboratories. The total number of hours instruction per year was 1,060 hours. Each school had a science room or laboratory. School sizes were the highest in the study: an average of 2,068 students. The pupil-teacher ratio was correspondingly high at 51.8. The curriculum is nationally prescribed by the ministry with inputs and collaboration of the teachers.

Population 2 was Grade 9 in Korea. It is at Grade 9 that science is split into separate subjects. The students came from large families. The students took science for 6.5 hours per week and did 2.3 hours of science homework per week. Their science teachers had an average age of 35 years and had had 10 years of teaching experience. Sixty-seven percent of them were male. They had had four years of postsecondary education of which 63 percent of the time had involved science. The inservice training in science was an average of 3.2 days per year. The teaching load was 20 hours per week and nearly all of this was spent teaching science. The schools had an average size of 770 students. The average number of laboratories was 1.6 per school with one laboratory assistant per school. About half the time of science instruction involved practical work.

Eighty-three percent of an age group was enrolled in Grade 12, but only 38 percent was enrolled in academic schools. It was the 38 percent which formed Population 3. Thirty-eight percent studied biology, 37 percent chemistry and 14 percent physics. The average age of students was 17 years 11 months and they studied at least nine subjects. The average class size was 58, the highest for all countries. The biology and chemistry students studied these subjects for about four hours per week and the physics students received seven hours. The total average homework (2.3 hours per week) was the least among all countries and about one-third (biology and chemistry) to half (physics) of this was devoted to science homework. Eighty-four to 96 percent (physics) of the teachers were male and their average age was 38 years. They had received 4.5 years of post-secondary education of which two thirds to three quarters was spent studying science. They received two days of inservice training per year for science and science teaching. The average weekly teaching load was 18 hours and they reported spending about 15 hours per week on preparation and marking. Only 14 percent of the schools were mixed. Forty-eight percent were boys schools and 38 percent girls schools. The average school size was 772 students.

Science achievement at Population 1 level was very high, at Population 2 level average and at Population 3 level relatively low compared with other countries. However, high proportions of the age group are studying science subjects at the terminal secondary school level. What is surprising is the high achievement at Population 1 level given the large class sizes.

Summary of Results 147

The Netherlands

The Netherlands tested only at Population 2 level. This was Grade 9. The average age of ninth graders in the Netherlands was 15 years and six months. Even the Netherlands researchers were surprised at this since it was expected that the average age would be within the 14-year-old age range. It would appear that many older youth returned to school to complete this grade level thus increasing the average age. The average number of children per family was 2.8. The students did six hours of science homework per week out of an average of 8.4 hours of total homework. The science teachers had an average age of 37 years and nearly 12 years of teaching experience. Eighty-seven percent of them were men. They had an average of 5.3 years of postsecondary education of which three years were spent studying science. They received two days of inservice training per year in science teaching. The total teaching load was 21.4 hours per week and 80 percent of that time was spent teaching science. Twelve percent of the schools were single sex schools and the average size of schools was 612 students. The schools were well equipped with laboratories and laboratory assistants. In the Netherlands, specialization in biology, chemistry and physics starts at the beginning of secondary schooling.

The school system is highly differentiated involving several school types each with a slightly different curriculum. The achievement of the Netherlands was among the highest at this level with the exception of biology which was only average. However, as was to be expected, the difference among schools was high.

Nigeria

Nigeria participated in Population 1 and 2 testing only. Population 1 was Grade 6. Nigerian children enter school anywhere between the ages of six and eight years. The average age of Grade 6 was 12 years and one month. They came from relatively large families having an average of 127 books per home. They were in classes with an average size of 32. Seventy-four percent of the teachers were men and the average age of all teachers was 28 (the youngest of all countries). They had had one year of postsecondary education of which only 11 percent was devoted to the teaching of science. Inservice training was only half a day per year. The teaching load was very low (14.3 hours per week) and the pupil-teacher ratio was 28.6. Science instruction was 2.5 hours per week. Twenty-six percent of the science instructional time was reported to be practical. However, 80 percent of the teachers perceived their science teaching as hampered by lack of equipment. The school curriculum is nationally prescribed. Eighty-six percent of the time, science was taught by the class teacher. Population 2 consisted of students in Grade 10. The average age was 16 years and two months. At this grade, science was taught as separate subjects. The exact percentage of an age group in school in Grade 10 was not reported. However, 63 percent of those in school were boys indicating a relatively large drop out of girls from the system. The students came from families of 4.5 children and this is probably an underestimate. They were in classes of 36. The students reported doing 7.4 hours science homework per week out of 10 hours of total time spent on homework. Their teachers had an average age of 33 years and had 9.5 years teaching experience. Eighty percent of the teachers were men. They had 4.8 years of postsecondary education and four fifths of that time had been spent studying science. They also received an average of 1.5 days per year inservice training in science. The total teaching load was low (13.7 hours per week) but most of that time (83%) was spent teaching science. Forty-eight percent of the schools were single sex schools and the average school size was 813 students. The schools seemed to be relatively well equipped with laboratories and laboratory assistants. Laboratories were used for about 53 percent of the time available. Achievement at both population levels was amongst the lowest in all aspects of science. Indeed, the average achievement at Grade 6 was the lowest of all countries and even Grade 10 achievement was lower than Population 1 achievement (i.e. Grade 4 or 5) in several developed countries.

Norway

Population 1 consisted of students in Grade 4. In Norway, children enter school at the age of seven years. The fourth graders had an average age of 10 years 11 months. They were from small families (2.2 children) and from 'literate' homes. They were in average class sizes of 20 students. Sixty-four percent of the teachers were female. Teachers had nearly four years of postsecondary education of which only eight percent was devoted to science. On average, the teachers were 41 years old with 15 years of experience. Nothing is known about their teaching of science nor about the schools in which they worked because Norway opted not to ask these questions. Norway tested Grade 9 instead of Grade 8 where most 14-year-olds were to be found. However, Grade 9 was the last grade of compulsory schooling and of the nine year comprehensive school. Hence, the average age was 15 years 10 months. The average number of children per family was three and the families had an average of 200 books. The students were in average class sizes of 23. There was no information on the number of allocated hours of science instruction. The teachers had an average age of 36 years and 51 percent of them were men. They had 4.7 years postsecondary education and 75 percent of that time was spent studying science. They received slightly less than a day a year inservice training in science. There was an average of 1.6 laboratories per school. Population 3 consisted of students in Grade 12 at which level 40 percent of an age group was enrolled. Four percent of an age group studied biology, six percent chemistry, and 10 percent physics. The average age of the students was 18 years 11 months. The average number of subjects studied was seven. The average school size was 377, and that only 68 percent of the teaching force was male.

At Population 1 level Norway was average compared with other countries. Its chemistry scores were relatively high but the biology and physics scores were low. At Population 2, it was also average with its chemistry scores being better than its other scores. At Population 3 it was again average but this time with its physics scores being relatively higher than its other scores. This is the first time that Norway has participated in an international study. The fact that its Grade 9 scores are lower than Sweden's Grade 8 scores may give it food for thought.

Papua New Guinea

Papua New Guinea participated in Population 2 and 3 only. However, at Population 3 level it did not take the specialist biology, chemistry and physics tests. It took a general science test (the results of which are reported in Appendix D). Since it was not included in the Population 3 comparisons no comments are made on Population 3 here.

However, for Population 2 Papua New Guinea tested Grade 10 where only 11 percent of an age group was enrolled in school (an élite group). Sixty three percent of Population 2 were boys indicating a relatively high drop out of girls before this point in school. Their average age was 17 years. The average number of children per family was 4.6 and this was probably an underestimate. Students reported their homes had an average of about 115 books. There were three hours per week allocated to science and the students reported doing 7.1 hours science homework per week out of 7.9 hours per week of all homework. The teachers had an average age of 40 years with nearly 15 years teaching experience. Seventy-five percent of the science teachers were men. They had a total teaching load of 18.2 hours and spent 71 percent of their time teaching science. Fifteen percent of the schools were single sex and the average school size was 513. The schools were reasonably well equipped: 1.9 laboratories per school and 1.6 laboratory assistants per school. Laboratories were used 75 percent of the time. This élite group, compared with other countries, scored in the middle range in biology but relatively poorly in the other subject areas.

Philippines

In the Philippines, children begin school at age seven. There are only ten grades of schooling. Science is taught in English which is not the mother tongue. The Philippines tested Populations 1 and 2 only. Population 1 consisted of all students in Grade 5. Their average age was 11 years and one month. They came from large families. The average class size was 40. Nearly all teachers were women with an average age of 41 years and 17 years of teaching experience. They had 4.5 years postsecondary education but it must be remembered that the Philippines ends secondary school with Grade 10. Only three percent of that postsecondary education was devoted to science. Three days a year were given to inservice training in science. The teaching load was, on average, 19.4 hours per week of which half were devoted to science. There was an even split between the specialist science and class teachers (23%) while in 47 percent of the schools, the class was taught by multiple teachers. Forty percent of the science time involved practical work but 63 percent of teachers perceived their science teaching to be hampered by lack of equipment. The total number of hours of instruction per year (1,359) was relatively high and the average size of school (1,204) was also high. The pupil-teacher ratio was 32.9. The curriculum was centrally prescribed at the national level by the Ministry of Education.

Population 2 consisted of all students in Grade 9, the penultimate year of all secondary schooling. Sixty percent of an age group was enrolled in school at this level. The students' average age was 16 years and one month. Science specialization began at Grade 8. Like the United States, biology, chemistry and physics were taught consecutively from Grade 8 to 10. Fifty-eight percent of the students were girls. The average number of children per family was 4.4 (comparable to Korea, Papua New Guinea, Thailand and Zimbabwe) and the average number of books at home was 78. There were five hours per week of allocated instructional time for science, and the students reported doing 4.7 hours science homework per week. Ninety percent of the teachers were women which was relatively constant across the primary grades. Their average age was 33 years and average teaching experience was 10 years. The teachers had had 4.8 years of post-secondary education but only .8 years of this was spent studying science. The average total teaching load was 20.5 hours per week and about four fifths of this time was spent teaching science. Only three percent of schools were single sex but the schools had relatively large enrolments (average size of 1,599 students). The schools seemed to be well equipped for science teaching. Those without laboratories conducted experiments in classrooms about 54 percent of the time.

At both grade levels achievement was low compared with most other countries. Indeed, as with Nigeria and Zimbabwe, the Population 2 scores were lower than the Population 1 scores in several developed countries.

Poland

Polish children also began school at age seven years. For Population 1 the grade tested was Grade 4 and the students had an average age of 10 years 11 months. Eighty-five percent of the teachers were women. Teachers had an average of nearly four years postsecondary education of which nearly half was given to science. Inservice training is 2.9 days per year. The teaching load was 23 hours per week of which 13 were spent teaching science [again the concept of specialist teachers]. Only 21 percent of science teaching was devoted to practical work. The schools had an average size of 614 pupils and the pupil-teacher ratio was 19.1. The curriculum was nationally controlled and teachers only had a choice about equipment. However, 56 percent of them reported that their science teaching was hampered by lack of equipment.

Population 2 consisted of students in Grade 8. Grade 8 was the end of the eight years of primary and compulsory schooling. In Grades 4 to 8 biology and geography were taught as separate subjects. In Grade 6, physics was added to the curriculum as a separate subject and in Grades 7 and 8 chemistry was added as a separate subject. The class size was 25 and the students reported doing 5.2 hours per week science homework which was just under 50 percent of total homework. The number of students reporting doing experiments was unusually small (11 percent) and the percent of time (as reported by teachers) spent on practical work was low (average of 23 percent of time). Even so, the schools reported having an average of 2.5 laboratories or equipped classrooms per school. The teachers had an average age of 37 years and had 16 years of teaching experience. This apparent anomaly was because many teachers taught and studied at the same time. Seventy percent of all science teachers were women. They had 4.1 years of postsecondary education of which 2.4 years had been spent studying science. The total teaching load averaged 22.4 hours per week and 70 percent of that was spent teaching science. The average school size was 583 and all schools were coeducational.

Population 3 consisted of students studying science in Grade 12. Twenty-eight percent of an age group was enrolled in Grade 12. Nine percent of an age group studied each of the science subjects. The average age of the students was 18 years seven months. They studied more than nine subjects. A majority of the students were girls. They expected to have 3.5 years of further education. The average class size was 30. The biology and chemistry students studied their

Summary of Results

science subjects for 10 hours per week and the physics students for only 5.5 hours. They did about 11.5 hours of homework altogether of which roughly half was for science. Most teachers were female and aged about 40 years. All had had 5.5 years of postsecondary education of which three quarters had been devoted to the study of science. They receive an average of 3.4 days of inservice training per year. The teaching load for science teachers was relatively high (21 to 22 hours) and they reported spending another 14 to 16 hours per week on preparation and marking. About one quarter of all teaching time was allotted to practical work and 50 percent of lessons were taught in laboratories. All schools were mixed with an average size of 488. Seventy percent of all science teachers were female. There was an average of five laboratories per school and they were used 86 percent of the time.

At Population 1 level Poland's achievement scores were relatively low overall. In terms of the separate branches of science they were low in earth sciences and physics but average in biology, and chemistry. At Population 2 level, they were high in biology and chemistry, average in earth sciences and low in physics. At Population 3 level the situation was reversed where they were relatively good in physics, average in biology and low in chemistry. However, compared with other countries they rise as one ascends the school system.

Singapore

Since 1980, formal primary school science was introduced with effect from Grade 3. It was not introduced at Grades 1 or 2 because the study of two languages (English and mother tongue) needed more time at that stage. The Grade 5 (Population 1) students tested in this study would have studied science for two years and three months when they took the tests. They were, on average, aged 10 years and 10 months and were from families of three children. The homes had an average of 97 books. Seventy percent of the teachers were women. The teachers at Grade 5 had an average of 2.7 years of postsecondary education. Inservice training was an average of 1.4 days per year. The teaching load was 18.8 hours per week of which four were devoted to the teaching of science. In the teaching of science about 27 percent of science curriculum time is devoted to practical work. Only 13 percent perceived their teaching to be hampered by lack of equipment. This low proportion is due to the provision of adequate science activity areas in the classrooms. The average school size was 1,201 with an average class size of 37. Six percent of schools were for boys only and seven percent girls only; the rest were coeducational.

The Population 2 students were drawn from Grade 9 (or Secondary 3 level), their ages ranged from 13+ to 17+, the modal group being 15 years old. At the time of test administration, the students had completed four years of general science at primary level and two years at the lower secondary level. At Grade 9, differentiation between the humanities and sciences begins. Hence the percentage of science instructional time of all instructional time varied from of 7.5 to 20 percent. There was an even split between boys' and girls' enrolment as a whole, even though they came from single sex boys' (11%) or girls' (15%) schools. The class size was around 38 which was not very different from the primary level classes. The amount of science per week was about 5.0 hours. Schools were relatively well provided with science laboratories (about four per school) and they were used 74 percent of the time. About 48 percent of the time, science learning involved 'doing experiments' in the laboratories. There was still a predominance (64%) of female science teachers at Grade 9, with an average age of 35 years and about 10.7 years of teaching experience behind them. On average they had 3.6 years of postsecondary education. Of this, 77 percent of the time was devoted to the sciences. Their workload was quite similar to that at the primary level (17.7 hours per week). They were specialized in science and taught science about 79 percent of the time. Their workload was spent half the time in the science laboratories and the rest was spent on preparation, marking and teaching. The science curriculum is nationally prescribed but principals and teachers have full autonomy in the choice of textbooks and equipment to match the needs of their students.

The last year of schooling included all students either in Grade 12 (2-year junior colleges) or in Grade 13 (the 3-year pre-university centers). These made up 17 percent of the 18 year old age group. A process of sampling was adopted to ensure that about the same number of students were drawn for the biology, chemistry, and physics sub-populations. Physics students included only the students who offered 'A' level physics (but excluded those who also enrolled in biology and/or chemistry). However, biology students included students also enrolled in physics and/or chemistry. As such, the relative proportion of males and females enrolled in physics and biology was confounded. In general, Population 3 students were the élite who prepare themselves for competitive admission to the university. On average, they expected to have an additional four years of tertiary education. The total number of hours of instruction was about 1,270 per year which was close to that of Israel which has the highest number of hours of total instruction. On average, school size was 1,500 students, and the average class size was 24. Schooling was entirely coeducational. Schools were well equipped for science teaching, with an average of 7.3 laboratories per school and the teachers were assisted by about five laboratory technicians per school. Biology students had the most number of hours of science instruction (18.6 hours per week) when compared with chemistry (12.2 hours), and physics (6.6 hours). This is because the majority of students who opted for biology were likely to enrol for chemistry and physics as well. Most of the time, the students who enrolled for physics, tended to select a combination of one or two advanced mathematics subjects, in place of science. Characteristically for the junior colleges, all students were engaged frequently in the conduct of experiments in the laboratories (on average 58 percent of the time). The students reported that they spent about 15 hours on all homework. In proportion to the number of hours enrolled for the sciences, biology students devoted about 64 percent of their time to science homework, in contrast to an average of about 50 and 37 percent for chemistry and physics. The majority of science teachers were in their mid-thirties. Given three to four years of tertiary education they spent about 90 percent of the time specializing in one of the science subjects. While the chemistry and physics teachers had between seven to nine years teaching experience, the biology teachers were much older with about 13 years of experience. Correspondingly, the biology teachers received about 2.4 days per year of inservice training which is about three times more than that for chemistry teachers. The physics teachers reported an average of .8 day per year only of in-service training. When they taught science, they devoted about 39 (physics) to 46 percent (biology) of the time to doing practical work. Between 55 to 65 percent of the time, teaching was conducted in laboratories. Except for about 13 percent of physics teachers, the majority of the science teachers did not perceive any problem with the laboratory facilities. Forty-two percent of the science teachers were male. The science teachers carried a workload of 15 to 16 hours per week and reported an additional 20 to 25 hours, preparing lessons and marking students' work. The overall impression is that science at Population 3 is specialized and equally demanding on students' effort and teachers' work commitment.

At Population 1 Singapore's achievement was relatively low in all branches of science. At Population 2 level it was average in biology and low in the other branches. At Population 3 level, it was high in all branches. This is very similar to the patterns of England and Hong Kong, i.e. poor achievement in science in primary education but very high at the end of secondary school. The proportion of an age group enrolled in the various branches of science at Population 3 was, however, about half that of Hong Kong.

Sweden

In Sweden children enter school at the age of seven years. At Population 1 level, two grades were tested in Sweden: Grade 3 where the students had an average age of nine years and 10 months and at Grade 4, where the students had an average age of 10 years and 10 months. All children came from 'literate' families, with homes well stocked with books. In Grade 3, 97 percent of the teachers were women but in Grade 4 only 62 percent were women. The average ages of the teachers were the highest of all countries: 49 years and 47 respectively for Grades 3 and 4. The teachers had two years of postsecondary education and received 1.5 days per year inservice training. Nothing more can be said about the aspects of teachers teaching science or any school variables since Sweden decided not to administer all of these questions.

For Population 2 Sweden tested Grades 7 and 8 which were part of the nine-year comprehensive school. Their ages ranged between 14-15 years. The average number of children per family was 2.6 for both grades and the average number of books per family was over 200. The students did very little homework (about one hour per week) when compared with other countries. They received an average of 3.7 hours per week instruction in science. The science teachers had an average age of 45 years and three quarters of them were men. They had only two years postsecondary education and half a year of that time was spent studying science. They received on average two days a year inservice training. Their total teaching load was about 23 hours per week and 65 percent of that time was spent teaching science.

Population 3 consisted of students in the academic stream in Grade 12. Twenty-eight percent of the age group were enrolled in the academic stream. Five percent of an age group studied biology and six percent chemistry and 13 percent physics. Science was taught as separate subjects. The students reported that they devoted about 37 to 45 percent of the total homework time (about 10 hours per week) to science. Over 90 percent claimed that they conduct experiments as part of their science learning. Average class sizes ranged from 22 to 24 students. The number of weekly hours of study was 2.0 for biology and 3.8 for physics.

Sweden achieved among the high scoring countries at both Populations 1 and 2 and also had small differences among schools. At Population 3 level, it scored relatively well in biology and was average in physics.

Thailand

Thailand tested only Populations 2 and 3. Population 2 was Grade 9 and Population 3 Grade 12. Only 32 percent of an age group was enrolled in Thailand at Grade 9 level which is the last year of the Lower Secondary School. This is an élite group: the average age was 15 years four months. The average number of children was 4.3 and the homes had an average of 160 books. The fathers had an average of 10 years of education. The class size was about 42. The students received three hours of science instruction per week and did 7.9 hours total homework. Of this amount 3.6 hours were for science. Sixty-two percent reported doing experiments. About 50 percent of the science time was given to practical work. Sixty percent of science teaching was done in the laboratory. There was an average of 1.5 laboratories per school manned by one lab assistant. The science teachers had an average age of 28 years and 6.5 years teaching experience. There were 50 percent men and 50 percent women teachers. Their postsecondary education lasted four years and 2.5 years of that were spent studying science. The average number of days per year inservice training in science was 2.2 days. The total teaching load was low (15.3 hours per week) and nearly all of that was devoted to the teaching of science. Six percent of the schools were single sex and the average school size was 1,113 students.

In 1984, Thailand had 29 percent of an age group enrolled in school at Grade 12. Fourteen percent were in academic schools and 15 percent in vocational schools. Vocational schools were not tested in this study. In the academic schools, seven percent of an age group studied each of the science subjects. The average age of the students was 18 years three months and they studied six subjects. In all subject areas, just over 50 percent of the students were girls. They all expected to have three years tertiary education. They were in classes of 40 students. All reported doing experiments in their science classes. The students studied 10 hours of science per week, and did about 7.5 hours of homework per week of which 4.6 hours were for science homework. Thirty-five percent of biology teachers, 46 percent of chemistry teachers and 84 percent of physics teachers were men. Their average age was 32 years, and they had had 4.4 years of postsecondary education of which two thirds were spent studying science. They received three to 4.5 days of inservice training per year. The average teaching load was 13.5 hours per week which was the lowest of all countries. They reported spending an additional 15 hours on preparing lessons and marking students' work. Forty to 50 percent of their teaching was devoted to practical work and 60 to 70 percent of teaching was done in laboratories. About 20 percent of biology and chemistry teachers in contrast to 35 percent of physics teachers perceived the lack of equipment to hamper their teaching severely. Ninety-four percent of schools were mixed and six percent sex-segregated. Forty-one percent of the total teaching force was male while 47 percent of all science teachers were men.

At Population 2 level Thailand's overall achievement score was in the middle group of countries. These countries included Australia, England, Hong Kong, Israel, Norway, Poland, Singapore and Sweden (Grade 7). The Thai scores

Summary of Results

were relatively high in biology, but relatively low in the other branches. At Population 3 level, all scores were relatively low in comparison with other countries.

The United States of America

The United States tested Grades 5, 9, and 12 for the different population levels. The average age of the students at Grade 5 was 11 years three months. Seventy percent of the Grade 5 teachers were women. The teachers had had 3.3 years of postsecondary education of which 20 percent had been devoted to science. They received an average of 1.8 days per year inservice training. Their average age was 40 years and they had had about 15 years of teaching experience. Their teaching load was 24.5 hours per week of which nearly six hours were for science teaching in classes of 24 students. It is to be noted that only six percent of United States schools had 'specialist' primary science teachers. Sixty-two percent of science teaching was handled by the class teachers while about 24 percent was taught by multiple teachers. Twenty-six percent of the science teaching time was spent in laboratory work. The United States chose not to administer many of the questions on the background questionnaire and hence a lot of information was missing.

The average age of ninth graders in the United States at the time of testing was 15 years three months. The average class size was 22 which was the second lowest of all countries. The students received 4.4 hours of science instruction per week and reported doing 9.5 hours homework per week. The teachers had an average age of 38 years and 14 years of teaching experience. Sixty-four percent of science teachers were men. The science teachers had received an average of 3.4 years of postsecondary education of which 2.1 years had been spent studying science. They received an average of three days per year inservice training. Their total teaching load was about 24 hours per week and they spent 73 percent of this time teaching science. Only 19 percent of science instruction took place in laboratories despite the fact that the schools stated that they had an average of five laboratories or equipped classrooms per school. However, these laboratories were in use, on average, just under 80 percent of the time, although, on average, there was less than one laboratory assistant per school. Six percent of schools were single sex. The average school size was 1141 students.

Eighty-three percent of an age group was enrolled in Grade 12. The United States tested students who continued to study biology, chemistry, or physics courses for the second year in the senior high school. For biology, several of the students were in Grade 10 whereas for chemistry and physics some were in Grade 11 and many in Grade 12. The estimated percentages of those studying biology, chemistry, and physics of the equivalent age group were given as 12, two, and one respectively. It was clearly difficult, if not impossible, for the National Center for Educational Statistics, an organ of the U.S. Department of Education, to furnish accurate figures so that the figures given above should be regarded as 'guesstimates' rather than estimates produced from data files. Many of the students studying a second year of science had parents with 14 to 15 years of education thus making them the children from fairly élite families. The students themselves, also expected, on average, to have three years of postsecondary education. They spent 13 hours a week on homework of which three hours were for science homework. Nearly all reported doing experiments as part of their science study. Half of the teachers were men for biology and physics but for chemistry 60 percent of the teachers were men. The teachers had an average age of 43 years, with nearly six years of postsecondary education of which 70 percent was spent studying science. They also received three days of inservice training in science and science teaching per year. The average teaching load (21 to 22 hours) was one of the highest in the participating countries and the teachers reported preparing lessons and marking students' work for 17 to 19 hours per week. Over 80 percent of teaching was conducted in laboratories.

Schools at Grade 12 were well equipped with laboratories (5.9 per school) which were used 82 percent of the time and staffed by one to two laboratory technicians. Ninety-five percent of schools were mixed; four percent of schools were girls' schools and one percent boys' schools. The average school size was 1,205 students. Forty-nine percent of all teaching staff were men but 63 percent of all science teachers were men.

At Population 1, the United States achievement scores were in the middle group of countries such as Australia, Italy, Norway, and Sweden (Grade 3). At Population 2 level they were relatively low and similar to England, Hong Kong, Italy (Grade 8), Singapore and Thailand. At Grade 12, they were amongst the lowest in biology and chemistry and somewhat better in physics. The differences among schools were high presenting the general picture that the education of this élite group achieved neither quality nor equality.

Zimbabwe

Zimbabwe tested only Population 2 (Grade 9). Only 30 percent of an age group was enrolled in school at this point in the educational system. Their average age was 16 years two months. Sixty-four percent of the students were boys from large families (average 4.4 children) and with an average of only 59 books per family. The students were in classes of 39, doing 5.3 hours a week science homework (out of a total of 8.1 hours on all homework) and 91 percent of them said they did experiments. Their teachers had an average age of 31 years and had had just over seven years teaching experience. Seventy-three percent of the teachers were men. They had had 3.4 years of postsecondary education of which 1.8 years was spent studying science. They received two days per year inservice teacher training and taught 19 hours per week most of which was teaching science.

The achievement, as measured by the IEA tests, was low in all branches of science. Again, the Grade 9 achievement was lower than the Grade 5 achievement in most of the developed countries.

Part 2: Review of Major Findings

A system of education is responsible for the education of the cohort of children born each and every year. Nearly all of education takes place in schools. There are a few non-formal type educational programs which occur in homes in villages and there are a few distance school learning programs. Such programs

Summary of Results

were not included in this study since the definition of the target population was all children in main-stream schooling.

Systems typically divide education into cycles: primary education grades, lower secondary grades and upper secondary grades. However, the number of grades in a cycle varies both across systems and within systems. Some systems operate a 6:3:3 organization, others 3:3:3:3, others 6:3:4 and so on. There are many variations. The teacher training system will often have different types of teacher training for the different grade spans or cycles. The curriculum will vary especially where there are different school types at secondary school. How the science curriculum is structured varies both within and between countries. Some teach general science, others have biology, chemistry and physics as separate subjects; sometimes these are compulsory but sometimes optional. Some systems teach biology in one year and chemistry and physics in the next year.

Class sizes and school sizes vary. The availability and use of science equipment varies.

The cognitive achievement results have been summarized in Chapter 3 and will not be repeated here. The attitude results for Populations 2 and 3 have been reported in Chapter 4 and will not be reported here.

This short summary reports the findings of associations among input and process variables and science achievement. The study deals only with science. In some cases, it might be the case that some findings in science could be general to the system. But, this is a matter on which the educational specialists in a country might wish to speculate.

The overall findings are presented in clusters: homes; school facilities; curriculum organization, control and implementation; secondary school organization; teacher training; school organization; and national organization and expectations.

Homes

In nearly all countries 10-year-old and 14-year-old children from higher status homes and homes with more reading resources, and homes with parents with more years of education performed better on the science tests than children from poorer homes with respect to these characteristics. Even when compared with school organizational and teacher and teaching factors the home influences are still very strong in both developed and developing countries.

As has been commented on above, this is not surprising since the differences among homes are much greater than the differences among schools and teachers. There is no intellectual minimum entry qualification needed to be a parent. On the other hand, very bad schools are not allowed to exist and teachers must obtain a teaching qualification before they are allowed into schools.

Much has been written about what constitutes family encouragement for children's learning (see Kellaghan 1991). There are also examples of school activities to help poor families begin to take up some of the home behaviours which encourage children's learning. It is clear that having more encouraging parents is still a major area for improvement in all countries.

School Facilities

There are many aspects of school facilities without which, it is believed, learning will be difficult. These include the existence of desks and seats for the students, a blackboard, sufficient text books and reference books, sufficient lighting and so on. In this study, the information collection limited itself to the existence of science laboratories or well-equipped classrooms.

At the primary grades level there was no effect of science facilities on achievement, but at Population 2 and 3 levels there were several countries where the lack of facilities (particularly in smaller and rural schools) were associated with lower achievement. At the higher grade levels the percentage of time teachers taught their classes in laboratories and the extent to which students undertook experiments had an effect on achievement. Hence, it is important that in those countries where there is a lack of science facilities and where this is having an effect on science achievement, the further equipping of some schools is necessary. No detailed data were collected about the laboratories and, therefore, nothing can be said about the optimum equipment for laboratories. But, in those countries where the lack of science facilities has an effect, it is a matter of bringing the poorer equipped schools up to the standards of the better ones.

Curriculum Organization, Control, and Implementation

Some school systems are more centralized and some more decentralized in their administration and decision making. At one end of the continuum there are set national curricula and at the other end teachers can decide what to teach. As we have seen from Table 5.2 there is a greatly varied picture of locus of control for different aspects of the curriculum.

Whether a secondary school system is selective or not makes a difference in terms of the science curriculum. Where there are different types of schools (often having different types of students attending them as in the Netherlands) then there are different science curricula for the different school types. In comprehensive school systems the intended curriculum is often the same for all children. But, in some countries the 'same' curriculum is spelt out in detail and in others only in broad outline leaving a great deal of freedom for teachers to implement within the broad general guidelines.

However, where the curriculum is nationally set, decisions are sometimes taken concerning general science courses as opposed to science taught in the separate branches of biology, chemistry, and physics; or, on having science taught as general science to a particular point in the school system and then as separate branches.

Detailed analyses were conducted at the Population 2 level only. It was found that:
1. Decisions about what science was taught occurred either at a level above the school (i.e. district, state, or national) or at the level of the individual teacher. More centralized decision making about the curriculum resulted in fairly similar offerings across all schools; individual teacher decisions resulted in different offerings; split decisions where there was a mixture of regional and teacher decisions

Summary of Results

about the curriculum resulted in very great differences in offerings over schools.
2. The grade span organization of schools and within schools related strongly to the grades in which teachers taught; the higher the grades in which teachers taught the more likely it was that the branches of science were taught separately.
3. There was wide variation across countries in both the existence of general vs. subject specific science courses and the extent to which schools varied in their offerings of these courses.
4. Both the kind of courses offered and the extent to which such courses were compulsory had important effects on the opportunity the students had to learn the content of the tests.

Two more speculative findings are:
5. Compulsory courses were found to have positive effects on achievement in several countries.
6. Within countries separate but simultaneous courses in biology, chemistry, and physics often had positive effects on science achievement as compared with general science courses or only single subject instruction.

What, then, are the implications of these findings? Whatever the organization of the educational system at the secondary level, having compulsory rather than optional science courses and having separate science courses (e.g. biology, chemistry, and physics) would seem to be worthy of consideration by educational planners should they wish to raise the level of science achievement in their school system.

In terms of equity of achievement across all schools, a central or national curriculum which is standard for all schools ensures more equal opportunities to learn. Where there is no national curriculum and where the decision making is left to individual teachers then national teacher education training with equivalent content would appear to be one way of promoting similarity in the opportunity to learn important content.

Upper Secondary School Organization

Countries vary in how they organize secondary schools. In particular, they vary in the number of grades in the school system. In this science study there were school systems with 10 grades, or 11, or 12, or 13 grades. The percentage of an age group going through to the very end of secondary school varied from less than one percent to about 90 percent. The average number of subjects studied in the last grade of school varied from three to more than nine.

How Long Should Secondary School Be?

The number of grades in the school systems studied ranged from 10 to 13.
It was the Philippines which had 10 years of schooling. The Philippines only tested Grade 9 in secondary school and the achievement was poor. Sixty percent of an age group is enrolled and it does seem that the quality of the existing

secondary schooling (at least in science) should be improved before any extension of the system takes place. At the same time, it should be pointed out that in many developing countries the mean achievement scores may mask large differences between urban and rural sectors and between public and private schools.

It was Québec which had Grade 11 as the last grade of secondary schooling. Its achievement was certainly lower than other systems and it may wish to consider having 12 grades of secondary school. There was little difference in achievement between those having 12 and 13 grades and one wonders about the value added aspect of having a thirteenth grade.

Does More Mean Worse?

The cryptic sub-heading poses the question of whether allowing a higher percentage of an age group to proceed to the terminal grade of secondary school lowers the general level of achievement of the final year. Philosophies of educational systems differ on the number of students being allowed through to the final grade. Some systems such as Japan and the United States have very high percentages of an age group proceeding to the end of secondary school. Others have a relatively low percentage. It would appear that some regard continuation to the end of secondary school as either nearly a human right or a way of ensuring a better educated general work force, whilst others regard such a continuation as primarily for an élite preparing for tertiary education. In general, the higher the percentage of an age group proceeding to the final grade of education, the lower the mean level of achievement in science. This is true whether one takes all in school at the final grade or only those enrolled in science courses. Although this is a general trend it is quite clear that there are exceptions. A case in point is Hong Kong. At Forms 6 and 7, it has 27 and 20 percent of an age group enrolled in school. At Form 6 level, 20 percent of an age group studies chemistry and physics and have high achievement. Many countries have a lower percentage of an age group studying those subjects and they have lower achievement. Thus, although Hong Kong is an exception, it does point to what is possible and provides evidence for other countries to know what is possible.

When the achievement of the 'élite' of the final grade was examined there was no relationship between the percentage of an age group enrolled either in school in general or in science courses and the achievement of the élite.

Does the Number of Subjects Studied at the Final Grade of Schooling Make a Difference?

The average number of subjects studied at the final grade of secondary schooling ranged from three to nine (or more). No significant difference was found in achievement in biology, chemistry and physics, of the various systems of education in terms of the number of subjects studied. There are systems which have specialization in specific subject areas and those which do not. For example, it would appear to be the case in England that many students take only three subjects for 'A' level implying that they have taken three main subjects for two years. On the other hand, in Finland, Hungary, Korea, Poland and Sweden students continue to study nine or more academic subjects. In these countries, it is ensured that a mixture of arts and science subjects are studied. At the same time, some specialization in science subjects can take place.

It would certainly seem that it is safe for most systems to have their students studying five or six subjects.

Teacher Training

It has already been seen that the form of teacher training required in a system is partly dependent on the grade span organization of the school system. This in turn has to do with the placement of teachers into schools and on the decision making about the curriculum in a decentralized system. However, in terms of the length of the preservice training of teachers in the samples there was not much variation within countries in the amount of training teachers at any one level receive. However, in some countries there was considerable variation in the amount of time during preservice training that was devoted to science education.

In some systems inservice training is compulsory and in other systems it is optional. Where it is compulsory some systems make a deliberate effort to ensure that the weaker teachers are trained first. Where inservice training is optional it is a common phenomenon that the better teachers attend the courses. Although there are no data in this study on the quality of inservice courses, it is known that they do vary from a day's lectures (but, dare one say it, where the objective of the course is, in reality, more social than cognitive) to several days where the content involves the teachers in practical work of direct relevance to their daily work.

Teacher training does affect the amount of time the teachers use practical work in the classroom which in turn influences achievement. This is an indirect effect. There is also a direct effect of teachers on achievement.

School Organization

School Size and Class Size

How schools are organized is, in part, a function of tradition and of national education legislation. Two of the aspects of school organization have to do with school size and class size.

The average size of school was not related to average country scores.

Within countries, class sizes tended to be very similar and no significant effect was found. However, it was of interest to note that in the primary grades the two highest scoring countries had the largest class sizes. These were Japan with an average of 38 students per class and Korea with 55 students per class. At Population 2 level, Japan (42 students) and Korea (65 students) were amongst the highest scoring countries. At Population 3 level England and Singapore had class sizes of 12 and 24 respectively. Unfortunately, no data were available for Hong Kong.

These data would seem to indicate that it is not class size *per se* that affects science achievement. Presumably those countries with larger class sizes include in their teacher training programs styles of class management to deal with larger classes. If Korea is taken as an example it would appear that they are successful. Hence, it is possible for planners to think of having larger classes but only on condition that teachers are trained in how to deal with such sizes.

Homework

Teachers in schools also set homework. The length of time spent on homework has a positive effect on achievement. Both homework in all subjects and homework in science are important. It is likely that the school culture, and work ethic determines how much homework is done. The teachers set the homework but at the lower levels of schooling home support is also important. The literacy of the home (number of books, use of dictionary) was influenced to some extent by the economic and educational status of the home. But, it was the literacy of the home which influenced the extent to which these students did their homework at home.

At the higher levels of schooling homework is undertaken by the students and the average hours of homework per week in science is relatively high.

National Organization and Expectations

Three aspects of general organization are reported here which are somewhat different from the secondary school organization touched above.

The first is equality of educational opportunity or, in other words, does it make a difference which school in the system a student attends? The second is the pace of learning which has to do with what content educational systems expect children to learn at which point in the system.

The third is the link between the primary cycle of education and the lower secondary cycle.

Does it Make a Difference Which School a Child Attends?

In some systems the achievement difference among schools at the primary level and among schools at the lower secondary level were very small and in other countries very high. It was in the Nordic countries and Japan where the achievement differences between schools were low. This means, for example, that it does not matter in Sweden whether a child goes to a school in Kiruna, Stockholm or Malmö, a rural school or an urban school. However, in the developing countries and some other countries it makes a lot of difference. The Nordic systems have achieved an equality of schooling in their schools. This has been a social educational aim in these countries for the last three or four decades and they have been successful. However, in many developing countries it is claimed that there is a large difference between private and public schools and between urban and rural schools and, hence, that children are advantaged or disadvantaged according to their parents' ability to pay for private schooling or according to where they live. In systems such as the Netherlands there is a deliberate policy of having many school types which are often streamed by ability

Summary of Results

and which have different amounts of science required. In such systems it makes a lot of difference in science achievement which school type a student attends. However, where there are large differences between schools and where the political aim is to have equality between schools, there is still much work to be done in terms of allocation of teachers to schools, school science equipment, and the curriculum offered and required.

'Pace' of Learning?

The pace of learning in this case means the amount of science learnt over a number of grades. It is quite clear that for a given number of grades in school some systems learn much more than others. The level of science achievement of fifth graders in some countries was higher than eighth or ninth graders in other countries. It would seem that the 'pace' of learning in some developing countries is half that of some developed countries. The reasons for this are not known, but there is no reason to believe that the students in one country should not learn as well as students in another country given the appropriate learning conditions. The lessons from this study are that the curriculum should demand more in the poorer achieving countries and that efforts should be made to ensure that the teachers are well trained in the subject matter knowledge.

The Relationship between Primary School Achievement and Secondary School Achievement

The correlation between the achievement of systems at the primary and lower secondary school levels was 0.84. This indicates that, in general, those systems performing relatively well or poorly in the primary grades also do so at the lower secondary grades. Although planners have to allocate resources across all levels of their school systems, it would appear that one priority for the poorer achieving systems in the primary grades is to pay special attention to science education at that level.

General Remarks

It would be presumptuous to suggest to the educational planners in each system what they should do. The major findings presented in this final chapter are based on *science* education and achievement *only*. Each country involved in the study will have also conducted its own national analyses. By examining international results what is possible can be seen by looking at countries with high achievement or of countries where there are small differences in achievement between schools, to cite but two examples.

But, different systems have different aims. Some wish to have differences between school types. It is up to national planners to examine their own data and to examine how other systems perform. There are political, social, and financial constraints within which planners operate. There is also the tradition of schooling and what teachers and parents are used to.

Thus, rather than make suggestions to planners about their own system a series of general questions are posed on which both planners and researchers might care to exercise their thinking.

Questions for National Planners

1. Does the present curriculum require enough of the students? Should all branches of science be offered and required as courses at the same time?
2. How should the teachers be trained in the subject matter and how should teachers' time be allocated for teaching within schools?
3. Does science education in the primary grades in the system need more emphasis and more time allocated to it?
4. Do science facilities need to be improved in some schools (or, in some cases, in many schools)?
5. How much homework should be given on a weekly basis? How can it be ensured that teachers mark all the homework and use the homework with the students?
6. How are science teachers kept 'up-to-date'? What types of in-service training are effective?
7. How many grades of secondary schooling should there be?
8. How many subjects should students study in the final years of schooling?
9. Should the percentage of an age group continuing past the end of compulsory education be expanded?

Questions for International Researchers

1. How should target populations be defined? Should variation be allowed?
2. How can the construction of international tests be improved to ensure comparability?
3. Is there ever a case for a total score being produced? Or does the different topic emphasis within different systems of education preclude this? If so, what kinds of achievement summaries should be used?
4. Should more effort be devoted to the measurement of the curricular delivery system. In particular, should more data be gathered on offerings, requirements, curricular sequence over grades, and teacher qualifications?
5. How can the training of National Research Coordinators (NRCs) be improved to ensure standard data collection and recording procedures?
6. How can the data cleaning phase be improved in terms of speed and quality?
7. How can the communication between the international coordination and data processing center and the NRCs be improved?
8. How can the educational planners' concerns be taken into account in international studies? (In other words, should the main policy questions, and therefore independent variables, be left only to researchers?)
9. Should the NRC be primarily a researcher or a subject matter person?

None of the questions posed above are easy to answer. But, they are questions for which answers are needed.

Reference

1. Kellaghan, T. 1991 *Home Environments and School Learnings*. Jossey-Bass, San Francisco.

APPENDICES

Appendix A

Name and addresses of participating institutions, General Assembly members and National Research Coordinators involved in the Second IEA Science Study

Country	Institution	General Assembly Member	National Research Coordinator
Australia	Australian Council for Educational Research P.O. Box 210 Hawthorn Victoria 3122	J.P. Keeves (1981-84) B. McGaw (1985-89)	M.J. Rosier
Canada (Eng.)	Ontario Institute for Studies in Education 252 Bloor Street West Toronto, Ontario M5S 1V6	B.J. Shapiro (1981-87) M.J. Connelly (1987-89)	M.J. Connelly (for central provinces) R.J. Crocker (for eastern provinces) H. Kass (for western provinces)
Canada (Fr.)	Departement des science de l'éducation P.O. Box 1250, Succ B. Université du Québec /Hull Hull, Québec J8X 3X7	G. Dussault (1984-88) (Science Study only)[1]	G. Dussault
China	Central Institute of Educational Research Bei San Huan Zhong Lu 46 Beitaipingzhuan Beijing	Teng Chun	Wang Shiqing
England	National Foundation for Educational Research in England and Wales The Mere, Upton Park, Slough, Berkshire SL1 2DQ	C. Burstall	W. Keys
Finland	Institute for Educational Research University of Jyväskylä Yliopistonkatu 9 40100 Jyväskylä 10	K. Leimu	K. Leimu
Ghana	Curriculum Research and Development Division Ghana Education Service P.O. Box 2379 Accra	R. Ntumi (1982-86)	R. Ntumi (1982-89)
Hong Kong	The Department of Education University of Hong Kong Hong Kong	M.A. Brimer (1981-87) J.B. Biggs (1987-89)	J.B. Holbrook

---

[1] The representatives indicated as Science Study only were from institutes which were members of the Science Study only and not of the full IEA General Assembly.

Country	Institution	General Assembly Member	National Research Coordinator
Hungary	National Institute for Education Berzsenyi u. 6 1087 Budapest	Z. Bathory	P. Vari
Israel	School of Education Tel Aviv University Ramat-Aviv 69978 Tel Aviv	D. Nevo	P. Tamir
Italy	European Center for Education Villa Falconieri 00044 Frascati	A. Visalberghi	M. Fierli
Japan	National Institute for Educational Research 6-5-22 Shimomeguro Meguro-ku Tokyo 153	H. Kida (1981-86) I. Suzuki (1986-89)	M. Miyake
Korea	Korean Educational Development Institute 20-1 Umyeon-Dong San Gangnam-Gu Seoul	Woong Sun Hong (1981-83) Young Shik Kim (1983-88) Se Hoshin (1988-89)	Soon Taek Kim (1981-83) Chang Jin Byun (1984-85) In-jae Im (1985-88) Jean-jae Lee (1988-89)
Netherlands	Department of Education University of Twente Postbus 217 7500 AE Enschede	E. Warries (1981-86) T. Plomp (1986-89)	H. Pelgrum
Nigeria	International Center for Educational Evaluation Institute of Education University of Ibadan Ibadan	E.A. Yoloye	S.T. Bajah
Norway	International Learning Cooperative Rosenhof skole Dynekilgt 10 0569 Oslo 5	P. Dalin (Science Study only)	S. Sjöberg (1983-86) A. Isnes (1986-89)
Papua New Guinea	Faculty of Education University of Papua New Guinea P.O. Box 320 Port Moresby	M. Wilson (Science Study only)	M. Wilson
Philippines	Institute for Science and Mathematics Education Development University of the Philippines Vidal A. Tan Hall Pardo de Tavera Street Diliman Quezon City 3004	D.F. Hernandez (1982-86) P. Jesuitas (1986-89)	D.F. Hernandez (1982-86) J.C. Fonacier (1982-89) V.M. Talisayon (1986-89) (Co-coordinators)
Poland	Institute of Teacher Training ul. Mokotowska 16/20 00561 Warszawa	B. Niemierko	K. Czupial E. Gabryelski
Singapore	Research and Testing Division Ministry of Education Kay Siang Road Singapore 1024	Phua Swee Leang (1982-83) Sim Wong Kooi (1983-85) Soon Teck Wong (1985-87) John Yip (1987-89) Kam Kum Wone (1989)	Yeoh Oon Chye Tan Yap Kwang
Sweden	Institute of International Education University of Stockholm 106 91 Stockholm	T. Husén (1981-83) I. Marklund (1983-89)	R. Noonan (1981-87) I. Fägerlind

Names and Addresses of Participating Institutions

Country	Institution	General Assembly Member	National Research Coordinator
Thailand	The Office of the National Education Commission Sukhothai Road Bangkok 10300	P. Sapianchai (1981-87) Panom (1987-89)	P. Soydhurum
U.S.A.	Teachers College Columbia University New York, NY 10027	R.M. Wolf	W. Jacobson R.Doran
Zimbabwe	Department of Curricular Studies University of Zimbabwe Mount Pleasant Harare	P. Gilbert (1981-84) (Science Study only)	P. Gilbert (1981-84) A. Dock (1985) L. Nyagura (1988-89)

Second IEA Science Study

Chairman   J. P. Keeves
International Coordinator   M. J. Rosier (1980-1988)
Chairman Science Study Committee   W. Jacobson

Persons Working on the Second IEA Science Study at the International Centers for More Than Six Months

J.P. Keeves	Australia	1980-1984
	Stockholm	1987-1989
M.J. Rosier	Australia	1980-1988
C.M. Kay	Australia	1982-1983
D. Couper	Australia	1980-1981
Heather Payne	Australia	1985-1987
S.B. Thoradeniya	Australia	1983
A. Wilson	Australia	1987
N. Sellin	Stockholm	1987-1988
L. Andersson	Stockholm	1988-1989
S. Böttcher	Hamburg	1988-1989
D. Jungnickel	Hamburg	1987-1989
D. Kotte	Hamburg	1987-1989
R. Lehmann	Hamburg	1987-1989
C. Morgenstern	Hamburg	1988-1989
A. Schleicher	Hamburg	1987-1989

Second IEA Science Study Steering Committee

Members:
    Z. Bathory (Hungary)
    W. Jacobson (United States)
    J.P. Keeves (Australia)
    S. Kojima (Japan)
    M.J. Rosier (Australia)
    P. Soydhurum (Thailand) (1984-89)

Representative of IEA-Standing Committee: K. Leimu (Finland) (1985-89)

Appendix B

Definitions of National Target Populations and Sampling

All statements in this appendix refer to 1984 or the year in which testing took place.

Part I: Definitions of National Target Populations

Population 1

Australia
All students of age 10:0 to 10:11 years in normal schools in Years 4, 5 and 6 on 31 October, 1983. Students for special schools such as for the handicapped were excluded from the target population. The excluded population was about 1.1 percent of the total population.

Canada (English)
All students in Grade 5 on 1 May, 1984 in all publicly supported schools where science was taught in English. Students in privately supported schools were excluded. The excluded students represented approximately 7 percent of the target population.

Canada (French)
All francophone students in Grade 5 on 15 May, 1984 in regular schools throughout Canada, where science was taught in French. It is to be noted that students in military schools and language immersion schools were excluded.

England
All students in normal schools in Year 5 in the age range 10:0 to 10:11 on 1 September, 1984; that is, of age 9:9 to 10:8 on 1 June, 1984. Students in special schools, at immigrant centers, in remedial classes in normal schools, and students in small schools with less than five students in the required age range were excluded. The excluded students represented about 1 percent of the target population.

Finland
All students attending Grade 4 classes in regular Finnish-speaking comprehensive schools on 15 April, 1984. Students from Swedish-speaking schools (5.8 percent of the population), plus teacher training schools, old-system secondary schools and all special education schools (3.2 percent) were excluded. The excluded proportion was 9.0 percent of all students in the population.

Hong Kong
All students in the fourth year of primary education in Hong Kong primary schools on 31 May, 1984. The students excluded were from schools following curricula linked to overseas countries such as the English Schools Foundation, and schools with composite grades. The excluded population was 2.7 percent of the target population.

Hungary
All students in Grade 4 on 20 May, 1983. Students in non-graded schools (2.9 percent) and in special education schools (3.3 percent) were excluded. The excluded students represented 6.2 percent of the target population.

Israel
All students in Grade 5 on 1 May, 1983 in all the schools where science was taught in Hebrew. Students in special schools were excluded. The excluded population was 2.5 percent of the population in Hebrew-speaking schools. The Arab schools were also excluded.

Italy
All students enrolled in the fifth grade of state elementary schools on 20 April, 1983. Students in private schools were excluded and they represented 7.7 percent of the target population.

Japan
All students in Grade 5 of primary schools on 15 May, 1983. Students in private schools and schools for the handicapped were excluded. This was 2.0 percent of the target population.

Korea
All students in the fifth grade on 11 November, 1983. Students in special schools and in schools in remote areas were excluded. These represented about 1 percent of the target population.

Nigeria
All students in the sixth year of a six-year primary school owned and run by the state in 1983. A three stage sampling procedure was used with states being selected first, then schools within states and finally students within schools. Eleven out of 19 states were selected at the first stage of sampling. Students in non-state schools represented approximately 10 percent of the target population and were excluded from the study.

Norway
All students of age 10:4 to 11:3 years in Grade 4 of normal schools on 6 April, 1984. The excluded students represented 1 percent of the target population.

Philippines
All students in Grade 5 classes in regular schools on 1 February, 1984. Students in special education schools and in schools with a total enrolment of fewer than 50 students were omitted. This excluded population represented 4.5 percent of the target population.

Poland
All students in Grade 4 of regular elementary schools with Polish as the official language on 8 May, 1984. Students in special schools were excluded and represented 0.9 percent of the target population.

Singapore
All students (age 9 to 11 years plus) in Primary 5 Normal and Primary 5 Extended Courses on 1 April, 1984. These schools offer two languages (English and mother tongue) and the same school curricula except that the Normal Course students complete their primary education after six years while the Extended Course pupils take an extra two years. Students in the Primary 5 Monolingual Course take only one language and do not take science; they were excluded from the study and represented 6.3 percent of the target population.

Sweden
Two separate grade levels were tested in Sweden:
1a: All students in Grade 3 of the comprehensive school on 15 May, 1983.
1b: All students in Grade 4 of the comprehensive school on 15 May, 1983.
Students in special education schools were excluded and represented about 1 percent of the age group.

United States of America
Phase 1: All students in Grade 5 classes of normal schools in the 50 states on 30 April, 1983.
Phase 2: All students in Grade 5 classes of normal schools in the 50 states on 30 April, 1986.

Population 2

Australia
All students of age 14:0 to 14:11 years in normal schools in Years 8, 9 and 10 on 30 September, 1983. Students in special schools were excluded. The excluded population amounted to 2.1 percent of the target population.

Canada (English)
All students in Grade 9 on 1 May, 1984 in publicly supported schools, where science was taught in English. The excluded students represented approximately 7 percent of the target population.

Definitions of National Target Populations and Sampling

Canada (French)
The francophone students in Grade 9 classes of the main stream in the regular schools throughout Canada on 15 April, 1986, where science was taught in French. It is to be noted that students in vocational streams, in special education schools, in military schools and language immersion schools were excluded.

China
All students in junior Grade 3 in key and ordinary schools of general secondary education in the metropolitan areas and their surroundings in Beijing, Tianjin and Taiyuan on 10 May, 1985.

England
All students in normal schools in Year 9 in the age range 14:0 to 14:11 on 1 September, 1984; that is, of age 13:9 to 14:8 on 1 June 1984. Students in special schools, at immigrant centres, in remedial classes in normal schools and in small schools with less than 25 students in the required age range were excluded. The excluded students represented 1 to 2 percent of the target population.

Finland
All students attending Grade 8 classes in regular Finnish-speaking comprehensive schools on 15 April, 1984. Students from Swedish-speaking schools (5.6 percent of the population), plus teacher training schools, old-system secondary schools and all special education schools (6.1 percent) were excluded. The excluded proportion was 11.7 percent of the population of students.

Ghana
All students in Grade 9 in traditional secondary, secondary technical, and junior secondary schools within the public education system on 10 May, 1984. Students in private and special schools (representing 2 percent of the age group) were excluded. In addition, students in middle schools (representing 35 percent of an age group) were not tested. The sampled students represented 6 percent of the age group.

Hong Kong
All students in the second year of secondary education (Grade 8) studying science as part of the Hong Kong Secondary School curriculum on 31 May, 1984. Students in schools offering overseas curricula such as the English Schools Foundation and special schools were excluded. The excluded population was 5.3 percent of the target population.

Hungary
All students in Grade 8 on 20 May, 1983. Students in non-graded schools (0.5 percent) and in special education schools (2.5 percent) were excluded. The excluded students represented 3.0 percent of the target population.

Israel
All students in Grade 9 on 1 May, 1983 in all the schools were science was taught in Hebrew. Students in special schools were excluded. The excluded population was 2.5 percent of the population in Hebrew-speaking schools. The Arab schools were also excluded from the testing program.

Italy
Italy tested two grade level populations at this level.
2a) Grade 8 is all students in Grade 8 (third grade of Scuola Media) on 20 April, 1983.
2b) Grade 9 is all students in Grade 9 (first grade of Superiore) on 20 April, 1983.
Whereas 99 percent of an age group was in school at Grade 8, only 72 percent was in school at Grade 9.

Japan
All students in Grade 3 of secondary schools on 15 May, 1983. Students in private schools and schools for the handicapped were excluded. The excluded students represented 3.5 percent of the target population.

Korea
All students in the 9th grade on 11 November, 1983. Students in special schools and in remote schools were excluded. The excluded students represented about 1 percent of the target population.

Netherlands
All students in Year 9 of Gymnasium, Atheneum, VWO (general secondary, preuniversity), HAVO (general secondary, senior), MAVO (general secondary, junior), LTO (lower technical vocational), LHNO (lower vocational: domestic science), LEAO (lower vocational: administrative and business), and LAO (lower vocational: agricultural) schools on 31 May, 1984. Students in schools which were various combinations of the above as well as students in the Middenschool, LNO and LMO were excluded. The excluded group represented about 2 percent of the population.

Nigeria
All students in Class 4 Secondary (Grade 10) of a five-year secondary school system owned and run by either federal or state governments in 1983. As in Population 1, a three-stage sampling procedure was used. Students in non-state schools represented about five percent of the target population and were excluded from the study. Furthermore, the 39 Federal Government Colleges were also excluded from the defined population.

Norway
All students of age 15:4 to 16:3 years in Grade 9 in normal schools on 6 April, 1984. Norway deliberately chose a year group that involved older students and was different from the IEA international definition. Grade 9 in Norway is the last grade of full-time compulsory schooling in Norway. The excluded students represented 1 percent of the target population.

Papua New Guinea
All students in Grade 10 in provincial high schools in October, 1984. Students in international schools and seminaries were excluded from the target population. Eleven percent of an age group is enrolled in school at Grade 10 level.

Philippines
All students in Grade 9 of regular schools on 1 February, 1984. Students in special education schools and in schools with a total enrolment of fewer than 50 students were excluded (24 percent of the target population). The target population represented 60 percent of the age group.

Poland
All students in Grade 8 of regular elementary schools with Polish as the official language on 10 May, 1984. Students in special schools were excluded. These students represented 0.9 percent of the target population.

Singapore
All students in Secondary 3 or Grade 9 (ages 14 to 16 years plus) on 26 April, 1984. They represented 91 percent of the age group. Students who pass the Primary School Leaving Examination are, on entry into the Secondary School, streamed into the Special, Express or Normal Courses. The Special and Express students are in the faster streams, taking four years to reach the GCE O-levels. The Special Course students are offered two first languages (Chinese and English) while the Express Course students are offered English as a first language and their mother tongue as a second language. Normal Course students follow a less demanding curriculum and take the GCE N-levels after four years. If they do well, they will proceed to the fifth year, leading to the GCE O-levels. At Secondary 3, science is an option on the school curriculum in contrast to two years of compulsory science taught at Secondary 1 and 2 levels.

Sweden
Two separate grade levels were tested at Population 2 level:
2a: All students in Grade 7 of the comprehensive school on 15 May, 1983.
2b: All students in Grade 8 of the comprehensive school on 15 May, 1983.
In both populations, students from special education schools were omitted and in both cases they represented about 1 percent of the target population.

Definitions of National Target Populations and Sampling

Thailand
All students in Year 9 in normal schools under the Department of General Education and the Office of the Private Education Commission on 30 December, 1983. Students in special schools and demonstration schools were omitted. These students represented about 1 percent of the target population. The target population represented 32 percent of an age group.

United States of America
Phase 1: All students in Grade 9 classes of normal schools in the 50 states on 30 April, 1983.
Phase 2: All students in Grade 9 classes of normal schools in the 50 states on 30 April, 1986.

Zimbabwe
All students enrolled in Grade 9 in May, 1983. There were 751 schools in the country having a ninth grade which formed the target population: 149 were government schools and 602 were private schools. It should be mentioned that children in Zimbabwe tend to enrol in school at 6, 7 or 8 years of age.

Populations 3B, 3C, 3P (Students Studying Science)

(See Table 1b for further information.)

Australia
All students in Year 12 on 1 August, 1983 who are studying one or more science subjects which meet the prerequisites for entry to tertiary science courses. Students in technical and further education colleges (TAFE) were omitted from the population; these students represented a small percent of an age group. The total target population of all students in Year 12 was 39 percent excluding students in TAFE colleges. Approximately 29 percent of an age group were studying science at Year 12 and approximately 10 percent were not studying science (see Population 3N). (Given the low response rate in 1983, the non-responding schools were invited to participate in a follow-up testing program in 1984).

Canada (English)
All students in Grade 13 in Ontario and in Grade 12 in all other provinces and territories in May, 1984 who were enrolled in science courses where science was taught in English. There are no Canadian data which gave the number of students enrolled in science in their final year since an unknown number of students took more than one science course.

Canada (French)
All francophone students of the main stream studying any science subject at the final year secondary level (the pre-tertiary level: Grade 11 in Quebec, Grade 13 in Ontario, and Grade 12 in all other provinces) in March, 1986, in the schools throughout Canada where science was taught in French at this level. It is to be noted that students in vocational streams, anglophone students, and those in military schools and language immersion schools were excluded. The percentage in an age group at the final year level was 79. In 3B it was 7, 3C-37, and 3P-35.

England
All students in the second year of an A-level course on 1 March, 1984 who were taking science subjects in schools. Students in Colleges of Further Education were excluded. (This group represented between 13 and 20 percent of all A level science students studying a science subject. However, from a research study by Keys (1986)[1] it would appear that the College of Further Education students did not have markedly different science achievement on the IEA tests from students in ordinary schools.) In 1984, about 20 percent of an age group was in school at this level. About half studied science and half did not.

---

1 Keys, Wendy (1986) "A Comparison of A-level Science Students in Schools, Sixth-Form Colleges and Colleges of Further Education", *Educational Research*, 28, (3), November 1986.

Finland
All students in the third grade of the Finnish-language general/academic upper secondary schools on 15 November, 1983. About 63 percent of an age group was in school at this year level, 41 percent was in upper secondary school and some 18 percent in vocational secondary schools. Vocational school students were not included in the target population. Of the 41 percent in academic schools, all took biology and about one third of these studied chemistry and/or physics.

Ghana
All students in traditional secondary and secondary technical schools in Grade 13 on 10 April, 1984, taking any three or four of the following subjects at A level: mathematics, physics, chemistry, biology. All regions of the country were included except for the Ashanti region. In other words, the regions included were: Greater Accra, Volta, Eastern, Western, Central, Brong Ahafo, Northern and Upper. At this level, about 1.2 percent of an age group was in school. About half of these students were enrolled in science courses. Hence, those studying science represented .5 percent (because the Ashanti region was not tested) of the age group.

Hong Kong
Two grade levels were tested in Hong Kong:
3a: Form 6 (Grade 12) consisted of all students in Anglo-Chinese and Chinese Middle School classes who were studying at least one science subject on 15 December, 1984.
3b: Form 7 (Grade 13) consisted of all students in matriculation classes who were studying at least one science subject on 15 December, 1984.
At Form 6 level, 27 percent of an age group was enrolled in school and at Form 7 level it was 20 percent. Of these, 12 percent at both levels studied at least one science subject. Those not studying science represented 15 percent and 8 percent respectively for the two levels; however, these students were not tested. Some Form 6 students (mainly Chinese Middle Schools) took a curriculum leading to an entrance qualification for the Chinese University of Hong Kong, although the situation has subsequently been made more complex by the introduction of a direct offer scheme by the Chinese University of Hong Kong in 1985 based on certificate level (Form 5) results.

Hungary
All students in Grade 12 academic schools in March, 1983, who were studying one or more science subjects for entry to tertiary education science courses. In Hungary, all students in the last two grades of academic schools were enrolled in a general science course. All students could then choose to enter for about 6 periods a week an elective course (which was an enrichment type course). Some chose to study physics, chemistry or biology. Thus, the students defined as belonging to Populations 3B, 3C and 3P were those who were taking the elective courses in these subjects. Those who were in the general science course only were designated Population 3N. Forty percent of an age group was in school at Grade 12 level. Twenty-two percent were in vocational schools and were not tested. Eighteen percent are in academic schools. Nine percent took the elective courses (Populations 3B, 3C and 3P) and 9 percent were in the general science course but did not take electives (Population 3N).

Israel
All students in Grade 12 who were enrolled in March, 1983 in a higher level science course taught in Hebrew (either biology, physics or chemistry) leading to matriculation examinations. (Given the low response rate in 1983, the non response schools were invited to participate in a follow-up testing program in March and April, 1984.)

Italy
All students studying any science subject in the final year of secondary school in all public and private schools. The final year is Grade 12 in the Istituti Magistrali and Grade 13 in the other types of schools. The other schools were Istituto Professionale per d'Agricoltura (3B, 3C), Istituto Teenico Agrario (3B,3C), Istituto Tecnico Industriale (3C, 3P), Liceo Scientifico (3C, 3P), Istituto Professionale per l'Industria e l'Artigianato (3P), Istituto Nautico (3P). There was no excluded population. Thirty-four percent of an age group was in school at this level. It was estimated that four percent studied biology, one percent chemistry and 13 percent physics.

Definitions of National Target Populations and Sampling

Japan
All students at the third year upper secondary school level studying Earth Science II (3E) and/or Biology II (3B) and/or Chemistry II (3C) and/or Physics II (3P) on 15 November, 1983. Students in vocational schools representing 31 percent of an age group were omitted. Sixty-three percent of an age group were in academic schools (Populations 3 and 3N). Twenty-eight percent of an age group was studying science and 35 percent was not studying science.

Korea
All students in Grade 12 studying any science subject on 1 April, 1983. It is to be noted that those students in special education schools and those in remote places were excluded. All of these excluded students, however, represented only 1.6 percent of the population. The percentage of an age group in school in Grade 12 was 83. This consisted of 38 percent of an age group in academic schools and 45 percent in vocational schools. Only students in academic schools were included in this testing program. The percentage of an age group studying science was 38 for 3B, 37 for 3C, and 14 for 3P.

Norway
All students in Grade 12 on 20 March, 1984 who were studying any science subject. In the sampling however, within one school only those studying one of the science subjects were tested. Thus, in school 001 only the biology students were tested, within school 002 only the chemistry students, within school 003 only the physics students, within school 004 only the biology students, and so on. A further complication was that only half of the students tested in a school were administered test 3M. The percentage of an age group in school (excluding vocational schools) in Grade 12 was 40, and the percentage of an age group studying was 4 for 3B, 6 for 3C and 10 for 3P.

Papua New Guinea
All students enrolled in the major science program (3 hours biology, 3 hours chemistry and 3 hours physics per week) in Grade 12 in the four national high schools and Kabiufa High School on 10 October, 1983. These students were defined as Population 3X and so took only tests 3M and 3X but not the separate biology, chemistry and physics tests. Only the performance of these students on test 3M is reported in this volume. The percentage of the age group in school at Grade 12 was 1.1 and the percentage in Population 3S in the major science program was 0.7.

Poland
All students in Year 12 in the mathematics/physics and biology/chemistry sections of general education schools on 10 May, 1984. Twenty-eight percent of an age group was enrolled in school at this level. Nine percent of an age group studied science.

Singapore
All students in the last year of schooling in the Junior College Year II classes and in the Pre-University Year III classes in the Pre-University Centres who were taking one or more science subjects at the GCE A-level on 26 April, 1984. They constituted 17 percent of an age group enrolled in school. Of these, 9 percent studied science. The students who majored in the sciences chose one or more from physics, chemistry, biology and physical science. Those who chose physical science were not permitted by regulation to take physics and/or chemistry, and vice versa. The sub-population 3P took A-level physics only. The 3C students took chemistry only or A-level physics and chemistry while the 3B students took biology but not physical science. For 3N a 10 percent random sample of non-science students were selected.

Sweden
All students in Year 12 in the science line (final grade of the three year course- Line N) and in the technology line (third year of either a three or four year course- Line T) in the upper secondary school on 15 May, 1983. Twenty-eight percent of age group was in school in Grade 12 and of these 13 percent were in Lines N and T. Five percent of the age group were enrolled in Line N and 8 percent in Line T. Students in the technology line did not normally study biology or chemistry. However, all students in the technology line were administered the chemistry test, although only approximately 10 percent of this group or 1 percent of the age group were studying chemistry. Thus, Population 3P represented 13 percent of the age group, 3B- 5 percent, and 3C- 6 percent.

Thailand

All students in the science group in Grade 12 in normal schools under the Department of General Education and the Office of the Private Education Commission on 30 December, 1983. Students in special schools, vocational schools and demonstration schools were omitted. In Population 3, as defined above, 14 percent of an age group is enrolled. Half of these study science (Populations 3B, 3C, 3P) and half do not (3N).

United States of America

Phase 1: All students enrolled in a physics course in Grade 12 of normal schools in the 50 states on 30 April, 1983. The United States tested only physics students (tests 3M and 3P in Phase 1 testing.) The physics students represented 13 percent of an age group. (Given the low response rate in 1983, the non-response schools were invited to participate in a follow-up testing program in 1984).

Phase 2: All students enrolled in second year biology (3B), second year chemistry (3C) or second year physics (3P) on 30 April, 1986. These students are spread across Grades 10, 11 and 12. It is to be noted that, in the 1986 testing, the core test (3M) was not administered and 5 items from the 3B test, 5 items from the 3C test and 4 items from the 3P tests were omitted. Furthermore, half of the attitude statements, 16 of the descriptive items, 10 of the word knowledge items and 8 of the math items were omitted.

Population 3N (Non-Scientists)

(See Table 1.2 for further information.)

Australia

All students in Year 12 on 1 August, 1983 not taking one or more science courses meeting the prerequisites for entry to tertiary science courses. The percentage of the age group in this population was 10.

Canada (French)

All francophone students of the main stream at the final-year secondary level (Grade 11 in Quebec, Grade 12 in all other provinces) in totally or partially francophone schools throughout Canada in May, 1986, not studying any science subject. Vocational curricula students were omitted. The percentage of the age group in this population was 21 percent.

England

All students in the second year of an A-level course not planning to take any science subjects at A-level. Students in schools with fewer than 12 students in the age group and in Colleges of Further Education were excluded. The excluded students represented an unknown percent of the target population. The percentage of the age group in this 3N population was 10.

Hungary

All students in Grade 12 academic secondary schools who were not studying science for entry into tertiary education. However, the normal course which they took included some biology and physics. The percentage of an age group in this population was estimated at 9.

Israel

All students in Grade 12 who were enrolled in March, 1983 in non-science higher level courses which led to matriculation examinations.

Italy

All students in the final year of secondary school not studying any science subject. The school types included were: Liceo Artistico, Liceo Linguistico, Istituto d'Arte, Istituto Professionale per il Commercio, Istituto Professionale Alberghiero, Istituto Professionale Femminile, Istituto Tecnico Commerciale, Istituto Tecnico per Geometri, Istituto Tecnico per il Turismo, Istituto Tecnico per Periti Aziendali, and Istituto Tecnico Femminile. Ten percent of an age group were in this population. In all, 34 percent of an age group was in school in the final year of secondary school.

Definitions of National Target Populations and Sampling

Japan
All students in the final year of the full-time normal course in upper secondary school not studying any advanced science subject. The percentage of an age group in this population was 35 percent.

Norway
All students enrolled in Grade 12 but not studying any science subject at the time of testing. The percentage of an age group in school, excluding vocational school students, at Population 3 level was 40. The percentage of an age group in Population 3N was 24.

Papua New Guinea
All students enrolled in the minor science program (3 hours general science per week) or studying no science in the four national high schools. The percentage of an age group in school at Grade 12 was 1.1. The percentage of an age group in Population 3N was 0.4.

Singapore
All non-science students in Pre-University Year III or Junior College Year II. These students represented 8 percent of an age group.

Sweden
All students in Year 12 in the upper secondary school not in the science or technology lines on 15 May, 1983. This group constitutes those students in economics (E Line), humanities (H Line) and social sciences (S Line). These students comprised 15 percent of the age group or a little over half of those enrolled in Grade 12.

Thailand
All science students in the non-science group in Grade 12 in normal schools under the Department of General Education and the Office of the Private Education Commission on 30 December, 1983. Students in special schools, demonstration schools and vocational schools were omitted. Seven percent of the age group was enrolled in Population 3N.

United States of America
Phase 1: All students in Grade 12 of normal schools on 30 April, 1983, and not enrolled in any science course. This group represented 66 percent of an age group.
Phase 2: Not tested.

Part II: Sampling

The International Sampling Referee proposed various sampling designs to the National Research Coordinators. Each National Research Coordinator prepared a sampling design for each target population to be tested. It was expected that the Sampling Referee would approve the sample designs before samples were drawn for testing. However, NRCs in Israel, Canada (English), Sweden and the United States (Phase 2) drew their samples without prior approval. Finland, Hungary, Italy, and Nigeria drew their samples on the basis of preliminary versions of their sample designs but before the final versions had been approved. The definitions of the target populations have been given above in Part I. Most National Research Coordinators conducted two-stage sampling with a probability proportional size (pps) sample of schools. The first stage was to draw schools with a probability proportional to the number of students in the target population in the school. Stratification was used to create homogeneous groups of schools typically by school type, size and geographical location. The second stage was to select a cluster (usually one intact class) of students at random within each selected school. Replacement schools were typically drawn by selecting the next school within the same stratum in the sampling frame.

The aim, in all cases, was to draw a sample which would yield a standard error of sampling of any estimated mean at the national level which was ± 5 percent of a standard deviation. In each case, the ratio of homogeneity (Roh) or intra-class correlation (Rho) was estimated by the NRC to determine the N of schools and students required.

However, there were some deviations from this general approach. Below, a brief summary is given of the ways in which the samples were drawn in each country. The response rates for schools and students can be seen in Appendices C1 to C7. The sampling errors for the national means on the core test are given in Appendix C8. For a detailed description of the sampling procedures see Rosier (1988)[1].

Australia
For all three populations, a two-stage pps sample was drawn. To permit comparisons between states within Australia, samples of about the same size were drawn from each state. This meant that different sampling fractions were employed for the eight state strata. At the first stage of sampling, schools were selected within strata with a probability proportional to the number of students in the target population in the particular school. At the second stage of sampling, a random cluster of 24 students within the target population was drawn (i.e. not an intact class). At Population 3 level the sample (drawn and tested in 1983) only had a 62 percent response rate. Since this was considered to be too small a percentage, a supplementary testing was carried out in 1984, and the response rate was raised to 79.7 percent.

Canada (English)
Canada (English) was divided into three regions: the Atlantic region, the Central region and the Western region. Within each region a simple random sample of schools was drawn but independent of size. Furthermore, oversampling of schools occurred in that the "designed" sample was 100 schools per region but 125 were approached to allow for anticipated losses. Selected schools were then stratified by province in order to establish province target population figures which allowed the calculation of stratum weights. Within each selected school an intact class was selected at random. The target population was established and school weights were calculated to correct for the differential sizes of schools. Stratum weights were then applied to correct for disproportionality among strata.

Canada (French)
For all three populations, two-stage pps sampling was used. The first stage of sampling was the selection of schools proportional to the number of students in the target population in each school. The second stage consisted of selecting at random 30 students (i.e. not intact classes) in the target population within each school for Populations 1 and 2, and 50 students for 3/3N (30 for Population 3 and 20 for Population 3N).

China
China tested Population 2 only. Two-stage pps sampling was used. At the first stage, schools were selected proportional to the number of students in the school. (In the sampling frame, schools were listed by district within municipalities.) At the second stage of sampling, students were selected at random (i.e. not an intact class) from within a school.

England
For all three populations, two-stage pps type sampling was undertaken. At the first stage, schools were stratified by school size. Schools were selected at random in each stratum. Because experience showed that 10, 20 and 40 percent refusal rates could be expected for Populations 1, 2 and 3 respectively, 11 percent (Pop 1), 25 percent (Pop 2) and 67 percent (Pop 3) more schools per stratum were drawn than in the designed sample. Over and above this, no replacement schools were drawn. At the second stage of sampling, a random sample of students (i.e. not an intact class) in the target population was selected. At Population 3 level, students were drawn at random for each subject (biology, chemistry, physics) from the list of students studying those subjects.

Finland
For all three populations, two-stage pps sampling was used. The first stage was a selection of schools proportional to the size of the target population in the school. At the second stage of sampling, one intact class was drawn at random. For Population 3, the sample schools administered no more than two specialist tests to its students, depending on a previously determined subject rotation scheme. This resulted in a sub-sample of the schools being tested in a particular subject. Two teaching groups were selected within each school, one representing students in the shorter

---

[1] Rosier, M.J. (1988) Sampling and Administration for the International Science Study. ERIC.

Definitions of National Target Populations and Sampling

mathematics course and the other the longer mathematics course. Regardless of the shorter/longer mathematics grouping, all students were requested to participate in the specialist subject tests which they were currently studying and to which the school was assigned. Also, all students took Test 3M.

Ghana
Ghana tested only Populations 2 and 3. Two-stage pps sampling was used. At the first stage schools were selected proportional to the size of the target population in each school. (Where a school was drawn and agreed to participate but was inaccessible by road, the next school on the sampling frame for that stratum was used.) At the second stage of sampling, a random sample of 24 students was drawn within schools. At Pop 3 level, students were divided into 3B, 3C and 3P.

Hong Kong
For Population 1, a probability sample of classes was drawn from a complete list of classes stratified by size of school and school session. School session defines the time of day in which the school operates. A proportional probability sample of classes was drawn from each stratum. All students in the class were tested. For Population 2, classes drawn were stratified by size of school, medium of instruction and whether the school was public or private. Medium of instruction refers to whether schools were Anglo-Chinese professing to use English as the medium of instruction or Chinese Middle Schools using Chinese (Cantonese). For Population 3, a probability sample of schools was drawn from the total stratified into types of schools. Stratification was by school type (government, aided, grant schools and private schools). Some strata included schools with only Form 6 students and other strata with Forms 6 and 7 students. Within schools, all students in the target population were tested with the biology, physics and chemistry tests. Science students not studying biology were thus required to take the biology tests. Their data were included in the results presented in the Preliminary Report (IEA, 1988) but have been excluded from this report.

Hungary
For Populations 1 and 2, a three-stage sample was selected. At the first stage, settlements (districts) on the basis of the number of 9- to 14-year olds living there were drawn. At the second stage, schools were selected on an equal probability basis from these settlements. Within the selected schools, one intact class was drawn at random. At Population 3 level, two-stage sampling was used. The first was a selection of schools with a probability proportional to size. The second stage was to select one intact class at random within each school.

Israel
At each population level a simple random sample of schools was drawn at the first stage of sampling. No stratification was used. At the second stage of sampling, one intact class was selected. Neither stratum target population statistics nor school target population statistics were available or could be made available. Hence, no weighting was possible. This is the only country where weighting was not undertaken. At Population 3 level all students studying science in the target population were included. For the non-scientists one class was selected. There was only a 33 percent response rate of schools in 1983 and in 1984 the other 67 percent were approached again. The designed N of schools was 125 and the final achieved N was 67.

Italy
At Population 1 level, schools were selected from strata (region and size of municipality within region) with equal probability. All students within the population in each selected school were tested. At Population 2 level, schools were selected with a probability proportional to size. There were separate sampling frames for Pop 2A and Pop 2B. The sampling frame for 2B was the same as for Pop 3. At the second stage of sampling, students selected at random within each school for Population 2 and all students within the target population were tested in Population 3.

Japan
For all three populations, two-stage probability sampling was used. The first stage was to draw a sample of schools proportional to the number of students in the target populations in that school. The second stage of sampling a cluster (one intact class) was selected at random from within each school.

Korea
For all three populations, two-stage probability sampling was used. At the first stage, schools were stratified by region (3) and size of school (3) and schools were then drawn at random with a constant interval. Within schools, 24 students were drawn at random (i.e. not intact classes) at Populations 1 and 2 levels. But at Population 3, the strata were region (3) by government/non-government by sex of school (male, female, mixed), i.e. 18 strata. The schools, and students within a school, were sampled in the same way as for Populations 1 and 2. Where a student was taking two science subjects he/she was randomly assigned to one of the appropriate subjects.

Netherlands
The Netherlands tested Population 2 only. A two-stage probability sample was drawn. At the first stage, schools were selected with a probability proportional to size of the target population in the school. At the second stage, an intact class was selected at random.

Nigeria
At both population levels, a three-stage probability sample was drawn. At the first step, the country was split into five zones (strata) and within zones, states were drawn. In Zone 1, Lagos, Oyo and Bendel; in Zone 2, Anambra and Cross River; in Zone 3, Kwara and Plateau; in Zone 4, Sokoto and Kaduna, and in Zone 5, Gongolo and Borio. In each state schools were subdivided into urban and rural schools. At the second stage of sampling, equal interval random sampling was used to select schools within urban and rural areas within states within zones. An intact class was then drawn at random within each school at the final stage of sampling. However, the population figures and school ID figures were given by superstrata (zones) only and it was at this level that the weighting was done.

Norway
For Populations 1 and 2, schools were first drawn with a probability proportional to size and at the second stage an intact class within each selected school was drawn. For Population 3 however, every school in the target population was randomly allocated a number from 1 to 214 (there being 214 schools on the sampling frame). The first school's science students took the biology test, the second the chemistry test, the third the physics test, the fourth the biology test, and so on. For Population 3N, every second school was taken and in these schools, the humanities and social science students were tested alternately. However, in the test administration, only half of the student sample within a school was given the IEA core test; the other half was given a national test instead.

Papua New Guinea
For Population 2, the sample was the total population of schools with all students being tested. After the data were returned from the schools to the National Center, a random 1 in 3 sample was made from each class for sending to the international center as the Papua New Guinea Population 2 file in order to reduce the costs of data coding and data entry. At Population 3 level, all students were tested and constitute the Papua New Guinea data set. No weighting was necessary.

Philippines
At Population 1 level, there were 15 strata: the thirteen regions of the Philippines plus two private school strata (those within the National Capital Region (Metro Manila) and those outside). Within each stratum, schools were selected with a probability proportional to the number of classes within the target population in each school. At the second stage of sampling, an intact class was selected using simple random sampling. At the Population 2 level, there were 39 strata (the 13 regions of the Philippines by three types of school: Barangay/Municipal High Schools, National/Provincial/City High Schools, and Private High Schools). The first and second stages of sampling were conducted in the same way as for Population 1.

Poland
For Populations 1, 2 and 3, the 49 administrative units (Voivodeships) of the educational system were grouped into 14 regions by town/village (28 strata). At the first stage of sampling, schools were selected within strata with a probability proportional to the number of students in the target population in the school. At the second stage of sampling, students were randomly selected from across classes (i.e. not intact classes). For Population 3, the strata were 14 regions by two types of

Definitions of National Target Populations and Sampling

school profile: biology/chemistry, and math/physics. Schools were selected within strata with a probability proportional to the size of the target population in each school. At the second stage of sampling, students were selected at random from across classes (i.e. not intact classes) according to the two profiles (tracks) in the school.

Singapore
As can be seen from Part I of this Appendix, Singapore streams students into different types of schools. For Population 1, there were five strata by type of school, and for Population 2, nine strata by type of school. At the first stage of sampling, schools were selected with a probability proportional to the number of students in the target population in the school. At the second stage of sampling, one intact class was selected at random and within each class, 24 students were selected at random. At Population 3 level, all students studying science were tested with the exception of one junior college which did not participate. Therefore, weighting was not applied.

Sweden
Municipalities were grouped into 29 areas (strata) according to:
a) type of economy: labor, capital, agricultural and mixed;
b) density of population: sparse, dense;
c) size of immigrant population: small, large; and,
d) rural/urban.

At the first stage of sampling, schools were drawn with a probability proportional to the number of students in the target population in the school. At the second stage of sampling, an intact class was drawn at random for each population tested. Thus, in Population 1, Sweden tested Grade 3 and Grade 4 and within the same school one Grade 3 class and one Grade 4 class were drawn. In this volume, these are reported as two separate Population 1 target populations. The same principles of sampling were used for the two Population 2 target populations (Grades 7 and 8) and for Population 3. However, problems were experienced with schools agreeing to participate and the school response rates were only 64, 60 and 68 percent for Populations 1, 2 and 3 respectively. No school replacement procedures were used.

Thailand
Thailand is divided into 13 educational administrative regions and these were the strata. At the first stage of sampling, schools were drawn at random with a probability proportional to the number of students in the target population. At the second stage, an intact class was selected at random within each school. At Population 3 level, one class was selected from the science track and one from the non-science track within each school. The ratio of science track to non-science track students is about 50:50.

United States
The United States conducted two phases of testing. Phase 1 and Phase 2.

Phase 1: For Phase 1, the first stage of sampling was a selection of schools by probability proportional to the school grade size from lists of public and non-public schools. At the second stage one intact class was selected within each school. No replacement schools were drawn. The number of schools required for any one of the population samples was doubled and all of these schools approached. Despite this, there was a relatively low response rate. An attempt was made to collect data in 1984 from the non-responding schools drawn in the 1983 sample. This was not successful. Hence, a new sample was drawn (Phase 2).

Phase 2: The sampling and field work for Phase 2 in 1986 was conducted by the Research Triangle Institute. The sampling plan for Populations 1 and 2 was a three-stage probability sample. The first stage was the sampling of counties or groups of counties proportional to the size of the the fifth grade target population. Each county having three or more schools was classified as one primary sampling unit. Counties having only one or two schools containing fifth grades were combined with other such counties within the same state in order to form a composite county PSU. A sample of 70 county PSUs was drawn. Sixteen strata were then constructed (public and non-public schools by metropolitan and non-metropolitan areas by four geographic regions: Northeast, South, North Central and Western). Metropolitan status was ascribed to a PSU if the PSU belonged to a standard metropolitan statistical area (SMSA).

Using this sampling frame, two schools were drawn in each selected county (primary sampling unit) with a probability proportional to size of the target population in the school. Within schools one intact class was selected at random.

At Population 3 level, the following procedure was adopted. RTI conducted an initial screening of the sample to ensure finding an adequate number of advanced chemistry and physics classes. Fifteen schools within each PSU which contained a twelfth grade were selected with a probability proportional to the estimated Grade 12 school enrolment. In several PSUs, less than 15 unique schools were available. The screening resulted in 738 schools of the maximum possible 1,050. Of these schools, 444 (60.2 percent) reported one or more advanced science courses and were willing to participate in the study. Further, 205 schools (27.8 percent) reported no advanced science courses and 89 schools (12.1 percent) declined to participate. Of the willing schools (444) 397 (89.4 percent) reported advanced biology courses, 245 (55.2 percent) reported advanced chemistry, and 105 schools (23.6 percent) reported advanced physics courses. Some schools reported more than one type of course (especially advanced biology) and a few reported more than one class of a given course.

From this information, RTI constructed three sampling frames; one each for biology, chemistry and physics. A list of schools within each PSU was constructed by affiliation (public, Catholic and other private) and by estimated grade enrolment. This sorting by affiliation and size imposed stratification of the school frame. The direction of grade enrolment sorting was switched at affiliation boundaries to provide for a smooth size transition. RTI selected 47 schools for advanced biology, 47 for advanced chemistry testing, and 46 schools for advanced physics testing. These selections were based on probability proportional to a weighted count of course selections with the limitation that a school would be selected only once for a specific advanced science course (e.g., biology). However, a given school could be selected for testing in two areas, such as biology and chemistry. Within each selected school, one intact class was selected.

It is to be noted that within schools it was advanced science classes which were tested at whatever grade level they existed. The following table presents the percentage of students at each grade level.

Course	Grade Level		
	10	11	12
Advanced biology	20	26	54
Advanced chemistry	0	33	67
Advanced physics	0	10	90

Hence, the United States departs from the international definition of the Population 3 students studying science in Grade 12. On the other hand, it can be argued that these were the students at the acme of their performance within the system.

In Phase 2, the United States collected the following information:

	Population 1	Population 2
Core test	yes	yes
Rotated test A	no	no
Rotated test B	no	no
Rotated test C	no	no
Rotated test D	no	no
Attitudes	5 items only out of 18	5 items only out of 40
Descriptive	5 items only out of 10	5 items only out of 40
Word knowledge test	no	no
Mathematics test	no	no
Opportunity to learn		
Percent	no	no
Grade	core only	core only

	Population 3
Core test	no
Earth science test	no

Definitions of National Target Populations and Sampling 185

Biology test	yes (25 international items out of 30)
Chemistry test	yes (25 international items out of 30)
Physics test	yes (26 international items out of 30)
General test	no
Nonspecialist test	no
Attitudes (56 items)	28 items only
Descriptive (40 items)	24 items only
Word knowledge test (40 items)	30 items only
Math test (20 items)	12 items only
Opportunity to learn	
Physics	yes (grade only)
Chemistry	yes (grade only)
Biology	yes (grade only)
Core	no

United States (Phase 2) comment
The representative of the National Science Foundation instructed the authors of this volume to use the Phase 2 data for all test scores reported in this volume. This was done. It is to be noted that in Phase 2 at Population 1 and 2 levels no rotated tests were administered. Furthermore, only a few of the attitude and descriptive measures' items were administered. At Population 3 level the biology and chemistry tests were administered in Phase 2; they had not been administered in Phase 1. Physics was repeated. But the United States dropped five items from the biology test, five from the chemistry test and four from the physics test on the grounds that the dropped items were not appropriate for the United States Finally, different measures of opportunity to learn were used on the two occasions.

Zimbabwe
Stratification was by region (6), government vs. private (2), and within government schools by three groups (all former white, all former black, and rural day secondary) and within private schools by three groups (high fee paying, medium fee paying, and private rural day secondary). Two stage probability sampling was used. At the first stage of sampling, schools were drawn with a probability proportional to the size of the target population and at the second stage 20 students were drawn at random from all students in Grade 9 (i.e. intact classes were not drawn).

Appendix C1

Number of Schools and Students in Population 1

Country	Schools					Students				
	Target Population	Designed Sample	Executed Sample	Achieved Sample	Response Rate	Target Population	Designed Sample	Executed Sample	Achieved Sample	Response Rate
Australia	7382	282	220	220	78.01	272891	6363	4259	4259	66.93
Canada (Eng.)	7212	314	215	215	68.47	255627	6751	5151	5104	67.20
Canada (Fr.)	2007	108	98	98	90.74	81496	3240	2752	2739	84.53
England	16460	275	181	181	65.82	586467	6048	3748	3748	61.97
Finland	3699	110	106	106	96.36	56274	1857	1631	1600	86.16
Hong Kong^a	2460	148	146	146	98.60	92218	5541	5352	5342	96.41
Hungary	3567	100	100	100	100.00	150130	2732	2596	2590	94.80
Israel^b	1110	96	86	86	89.58	60000	n a	2429	2351	n a
Italy	31227	205	119	119	58.05	937730	6133	5195	5156	84.07
Japan	20913	223	221	221	99.10	2008068	8000	7925	7924	99.05
Korea	5385	147	146	146	99.32	861868	3528	3489	3489	98.89
Nigeria^c	37465	112	78	78	69.64	2264706	2658	2152	944	35.52
Norway	3037	147	91	91	61.90	61358	2420	1329	1305	53.93
Philippines	30216	500	475	475	92.60	1122279	18247	16851	16851	92.35
Poland	14765	199	199	199	100.00	553719	4699	4390	4390	93.42
Singapore^a	378	251	222	222	92.43	42877	6024	5547	5547	92.08
Sweden (Grade 3)	1278	105	75	75	71.43	12239	1904	1397	1336	70.17
Sweden (Grade 4)	1278	92	64	64	69.57	113063	1960	1518	1449	73.93
U.S.A. (Phase 1)	-	253	121	121	47.83	3413376	6325	2009	2009	45.99
U.S.A. (Phase 2)	-	140	123	123	87.90	3153653	3665	2961	2622	77.00
Total	188639	3807	3096	3096		16200029	98095	83481	81855	

a   In Hong Kong and Singapore, the primary sampling units were classes rather than schools.

b   No student design sample figures were available. No target population figures were available and hence all data presented are unweighted.

c   At the requests of the Nigerian National Center students in the age range of 132-156 months were extracted from the executed sample to form the achieved sample.

na   not available

Appendix C2

Number of Schools and Students in Population 2

Country	Target Population	Schools Designed Sample	Schools Executed Sample	Schools Achieved Sample	Response Rate	Target Population	Students Designed Sample	Students Executed Sample	Students Achieved Sample	Response Rate
Australia	2144	276	233	233	84.4	246114	6624	4917	4917	74.2
Canada (Eng.)	2893	316	209	209	66.1	290032	6125	5639	5543	90.5
Canada (Fr.)	582	104	101	101	97.1	81666	3120	2405	2348	75.3
China[a]	1790	105	105	105	100.0	248241	2820	2817	2806	99.5
England	4358	247	147	147	59.5	708039	5928	3118	3118	52.6
Finland	598	93	90	90	96.8	61100	2820	2592	2546	90.3
Ghana	633	100	95	95	95.0	23949	2500	2769	2769	110.8
Hong Kong[b]	2166	133	132	132	99.2	85417	5244	4981	4973	94.8
Hungary	3567	100	99	99	99.0	140968	2704	2515	2515	93.0
Israel	543	100	74	74	74.0	53000	3000	2200	2082	69.4
Italy (Grade 8)	9349	298	224	224	75.2	898790	5960	4640	4622	77.6
Italy (Grade 9)	7341	104	72	72	69.2	375520	1856	1408	1398	75.3
Japan	9965	201	199	199	99.0	1819474	8000	7610	7610	95.1
Korea	2116	189	189	189	100.0	837462	4536	4522	4522	99.7
Netherlands	2759	244	224	224	91.8	257595	5856	5065	5025	85.8
Nigeria[c]	4476	224	82	82	36.6	441213	5376	2277	804	15.0
Norway	1103	118	77	77	65.3	66931	2429	1424	1420	58.5
Papua New Guinea	105	105	87	87	82.9	8445	2812	2272	2193	78.0
Philippines	4117	300	261	261	89.7	573110	12310	10888	10888	88.3
Poland	12607	201	201	201	100.0	488170	4749	4520	4520	95.2
Singapore	249	185	185	185	100.0	39158	4440	4430	4430	99.8
Sweden (Grade 7)	948	118	71	71	60.2	113351	2915	1649	1557	53.4
Sweden (Grade 8)	948	115	69	69	60.0	120694	2828	1590	1461	51.7
Thailand	2231	103	96	96	93.2	391780	4120	3780	3780	91.8
U.S.A. (Phase 1)	-	256	88	88	34.4	3621044	6400	1958	1958	30.6
U.S.A. (Phase 2)	-	140	119	119	85.0	2968305	3667	2614	2519	68.7
Zimbabwe	751	129	129	129	100.0	95009	2580	2648	2648	102.6
Total	78339	4604	3658	3658		15054567	121719	97248	94972	

a China tested only in Beijing, Tianjin, and Taiyuan (the provincial capital of Shan Xi).
b In Hong Kong and Singapore, the primary sampling units were classes rather than schools.
c At the requests of the Nigerian National Center students in the age range of 168+ months were extracted from the executed sample to form the achieved sample.

Appendix C3

Number of Schools and Students in Population 3 - All

Country	Schools					Students				
	Target Population	Designed Sample	Executed Sample	Achieved Sample	Response Rate	Target Population	Designed Sample	Executed Sample	Achieved Sample	Response Rate
Australia	1461	207	165	165	79.7	98688	7329	5057	5057	69.0
Canada (Eng.)	429	107	370	370	.	.	.	9925	9452	.
Canada (Fr.)	.	.	105	105	98.1	44693	3210	3619	3469	108.1
England	2736	258	127	127	49.2	149854	9174	3737	3737	40.7
Finland	411	92	86	86	93.5	33022	4033	3775	3638	90.2
Ghana	83	23	20	20	87.0	2080	600	495	494	82.3
Hong Kong (6)	272	163	158	158	96.9	12123	7303	6103	6025	82.5
Hong Kong (7)	194	117	115	115	97.4	6923	4185	3754	3701	87.9
Hungary	250	80	77	77	96.3	19860	2243	2019	2001	89.2
Israel	253	122	68	68	55.7	63000	4996	2530	1982	40.2
Italy	7309	457	317	317	69.4	392969	9140	6888	6848	74.9
Japan	35180	202	193	193	95.5	1280221	7200	6561	6561	91.1
Korea	758	210	210	210	100.0	331023	8400	8399	8333	99.2
Norway	214	214	165	165	77.1	9460	3210	1602	1597	49.8
Papua New Guinea	5	5	5	5	100.0	.	834	808	795	95.3
Poland	1121	150	150	150	100.0	40303	3600	3246	3246	90.2
Singapore	17	17	16	16	94.1	4107	4107	3263	3263	79.4
Sweden	275	175	119	119	68.0	13099	4200	2849	4033	61.9
Thailand	795	100	98	98	98.0	160556	4000	4716	4714	117.9
U.S.A. (Phase 1)	.	.	164	164	.	.	.	4774	4774	.
U.S.A. (Phase 2)	.	.	100	100	.	.	.	1729	1729	.
Total	50302	2492	2828	2828		2563293	80375	85849	85449	

Appendix C4
Number of Schools and Students in Population 3 - Biology

Country	Schools					Students				
	Target Population	Designed Sample	Executed Sample	Achieved Sample	Response Rate	Target Population	Designed Sample	Executed Sample	Achieved Sample	Response Rate
Australia	1461	198	164	164	82.8	45483	2276	1631	1631	71.7
Canada (Eng.)	1745	286	187	187	65.4	84860	6078	3407	3254	52.7
Canada (Fr.)	429	53	47	47	88.7	7988	340	256	249	73.2
England	.	.	123	123	.	.	.	884	884	.
Finland	411	46	43	43	93.5	33022	1975	1707	1652	83.7
Ghana	69	21	20	20	95.2	894	210	210	210	100.0
Hong Kong (6)	.	.	158	158	.	.	.	3008	2975	.
Hong Kong (7)	.	.	115	115	.	.	.	1579	1559	.
Hungary	.	.	71	71	.	.	.	304	301	.
Israel	.	75	54	54	73.0	4000	1470	1058	879	59.8
Italy	.	.	12	12	.	.	.	147	147	.
Japan	5762	40	38	38	95.0	194747	1347	1212	1212	90.0
Korea	.	.	207	207	.	328899	3429	3337	3319	96.8
Norway	.	.	52	52	.	.	.	277	276	.
Poland	.	71	71	71	100.0	19096	1704	764	764	44.8
Singapore	518	9	8	8	90.9	1070	1070	902	902	84.3
Sweden	9	.	119	73	.	.	.	657	619	.
Thailand	.	.	98	98	.	.	.	1171	1171	.
U.S.A. (Phase 2)	.	47	43	43	91.5	348328	858	659	659	76.8
Total	10404	846	1630	1584		1068387	20757	23170	22663	

Appendix C5

Number of Schools and Students in Population 3 - Chemistry

Country	Target Population	Schools Designed Sample	Executed Sample	Achieved Sample	Response Rate	Target Population	Students Designed Sample	Executed Sample	Achieved Sample	Response Rate
Australia	1461	199	164	164	82.4	30530	1520	1177	1177	77.4
Canada (Eng.)	1662	297	179	179	60.3	74760	5732	3108	2923	51.4
Canada (Fr.)	429	107	105	105	98.1	33999	1583	1245	1187	75.0
England	.	.	123	123	.	.	.	892	892	.
Finland	411	46	44	44	95.7	13209	1167	1001	971	83.2
Ghana	73	21	20	20	95.2	2024	210	350	350	166.7
Hong Kong (6)	.	.	158	158	.	.	.	6012	5952	.
Hong Kong (7)	.	.	115	115	.	.	.	3639	3599	.
Hungary	.	42	56	56	.	900	820	146	143	.
Israel	.	.	17	16	38.1	.	.	309	243	29.7
Italy	7209	43	24	24	100.0	268176	1584	217	217	92.7
Japan	.	.	43	43	100.0	319430	3313	1468	1468	89.9
Korea	.	.	205	205	.	.	.	2979	2979	.
Norway	.	.	46	46	.	19096	1704	284	283	44.9
Poland	518	71	71	71	100.0	.	.	765	765	.
Singapore	9	9	8	8	90.9	1270	1270	945	945	74.4
Sweden	.	.	119	119	.	.	.	1261	1172	.
Thailand	.	.	98	98	.	.	.	1169	1168	.
U.S.A. (Phase 2)	.	47	40	40	76.1	107470	771	537	537	69.6
Total	11772	882	1635	1634		870864	19674	27504	26971	

Appendix C6

Number of Schools and Students in Population 3 - Physics

Country	Schools					Students				
	Target Population	Designed Sample	Executed Sample	Achieved Sample	Response Rate	Target Population	Designed Sample	Executed Sample	Achieved Sample	Response Rate
Australia	1461	198	163	163	82.3	26679	1420	1073	1073	75.6
Canada (Eng.)	1544	281	181	181	64.4	57560	5221	2890	2766	53.6
Canada (Fr.)	429	101	99	99	98.0	30359	1269	969	944	74.4
England	.	125	125	125	.	.	.	917	917	.
Finland	411	46	42	42	91.3	11558	977	837	810	82.9
Ghana	73	21	20	20	95.2	2059	210	291	291	138.6
Hong Kong (6)	.	.	158	158	.	.	.	6077	6025	.
Hong Kong (7)	.	.	115	115	.	.	.	3730	3690	.
Hungary	.	75	75	75	.	.	.	400	398	.
Israel	.	.	39	36	48.0	1800	1046	588	472	45.1
Italy	5521	39	120	120	92.3	190490	1332	1773	1766	89.1
Japan	.	.	36	36	.	122028	826	1187	1187	118.8
Korea	.	.	172	172	100.0	21207	1896	444	443	.
Norway	.	79	55	55	90.9	1308	1308	981	981	90.5
Poland	603	9	79	79	.	.	.	1716	1716	81.9
Singapore	9	9	8	8	.	.	.	1071	1071	.
Sweden	.	.	119	119	.	.	.	1231	1156	.
Thailand	.	.	98	98	.	.	.	1169	1168	.
U.S.A. (Phase 1)	.	502	158	158	31.7	3086531	12550	2719	2719	21.7
U.S.A. (Phase 2)	.	46	35	35	76.1	28722	762	485	485	63.7
Total	10051	1397	1897	1894		3580301	28817	30548	30078	

Appendix C7

Number of Schools and Students in Population 3 - Non-Scientists

Country	Schools					Students				
	Target Population	Designed Sample	Executed Sample	Achieved Sample	Response Rate	Target Population	Designed Sample	Executed Sample	Achieved Sample	Response Rate
Australia	1461	200	120	120	60.0	98432	1907	995	995	52.2
Canada (Eng.)	.	.	81	81	.	.	.	520	509	.
Canada (Fr.)	429	92	89	89	96.7	18491	1716	945	889	51.8
England	.	.	126	126	.	.	.	1004	1004	.
Hungary	.	.	77	77	.	.	.	1046	1036	.
Israel	.	65	24	28	43.1	15000	1600	575	388	24.3
Italy	.	.	119	119	.	.	.	2476	2455	.
Japan	15556	62	60	60	96.8	589561	2347	2229	2229	95.0
Norway	222	111	78	78	70.3	13135	1665	597	595	35.7
Papua New Guinea	5	5	5	4	80.0	.	.	312	300	.
Singapore	.	.	9	9	.	.	405	297	297	73.3
Sweden	525	108	73	73	67.6	16523	2644	1362	1281	48.5
Thailand	.	.	98	98	.	.	.	1207	1207	.
U.S.A. (Phase 1)	.	502	134	134	26.7	3086531	12550	2055	2055	16.4
Total	18198	1145	1093	1096		3837673	24834	15620	15240	

Appendix C8

Marker Variable Check: Percent Male Students in Target Population and Sample

Country	Population 1			Population 2			Population 3 Biology			Chemistry			Physics		
	Popn	Sample	Standard Error	Popn	Sample	Standard Error	Popn	Sample	Standard Error	Popn	Sample	Standard Error	Popn	Sample	Standard Error
Australia	51	49.0	1.37	51	54.4	2.40	33	33.3	2.67	61	56.9	2.96	73	68.9	2.81
Canada (Eng.)	51	51.0	0.86	51	48.8	1.66	41	37.0	2.25	54	53.0	2.94	68	70.1	2.73
Canada (Fr.)	51	51.3	1.00	49	47.3	2.16	na	25.6	8.60	na	44.4	2.34	na	50.6	2.99
China	·	·	·	na	47.5	1.17	·	·	·	·	·	·	·	·	·
England	51	51.5	1.18	51	52.2	2.28	40	40.0	3.84	65	66.5	3.33	79	78.0	2.89
Finland	52	51.7	1.13	na	51.6	1.02	39	47.1	3.19	46	47.6	5.71	61	63.3	5.20
Ghana	·	·	·	69	70.4	2.90	na	81.8	6.83	na	87.9	5.46	na	89.1	5.18
Hong Kong	na	50.4	1.73	na	54.8	2.67	·	·	·	·	·	·	·	·	·
Hong Kong (Form 6)	·	·	·	·	·	·	·	·	·	·	·	·	·	·	·
Hong Kong (Form 7)	·	·	·	·	·	·	·	·	·	·	·	·	·	·	·
Hungary	51	50.1	1.15	51	49.2	1.00	na	68.1	2.07	na	75.8	1.42	na	76.5	1.39
Israel	50	48.9	1.67	51	46.4	5.37	na	68.3	3.25	na	76.9	2.04	na	77.5	2.00
Italy	52	52.7	0.79	na	46.4	5.37	40	35.6	4.20	na	36.2	6.87	na	71.2	2.79
Italy (Grade 8)	·	·	·	·	·	·	40	39.8	3.14	51	51.0	nc	71	68.7	6.16
Italy (Grade 9)	·	·	·	na	52.0	1.05	na	81.8	nc	na	69.4	nc	na	80.9	5.99
Japan	51	51.5	0.38	51	47.5	5.03	na	27.2	nc	na	73.6	nc	na	87.1	12.22
Korea	na	52.9	0.57	na	52.9	0.31	na	52.3	3.41	na	52.0	3.63	na	75.0	4.85
Netherlands	·	·	·	52	50.3	3.00	·	·	·	·	·	·	·	·	·
Nigeria	57	54.3	2.17	na	62.7	2.00	·	·	·	·	·	·	·	·	·
Norway	49	52.6	1.71	49	50.8	4.44	30	26.2	9.64	47	40.7	11.76	72	67.5	9.71
Papua New Guinea	·	·	·	66	63.2	1.11	·	·	·	·	·	·	·	·	·
Philippines	51	44.4	0.49	49	41.9	1.86	·	·	·	·	·	·	·	·	·
Poland	51	51.0	0.73	51	48.9	0.90	·	·	·	·	·	·	·	·	·
Singapore	51	50.3	1.81	50	50.9	0.82	24	24.1	3.78	24	24.0	3.90	46	46.0	3.46
Sweden	·	·	·	·	·	·	51	51.1	5.24	59	59.9	5.95	67	67.5	5.03
Sweden (Grades 3 + 7)	51	50.3	1.23	51	47.9	1.29	55	55.6	4.54	75	70.1	2.82	75	70.4	2.48
Sweden (Grades 4 + 8)	51	50.9	1.12	51	51.9	1.46	·	·	·	·	·	·	·	·	·
Thailand	·	·	·	50	47.6	2.45	47	45.9	2.59	47	45.9	2.68	47	48.3	2.62
U.S.A.	49	49.2	0.91	52	50.9	1.43	51	41.0	8.44	51	58.8	11.07	51	78.0	7.47
Zimbabwe	·	·	·	na	63.5	2.00	·	·	·	·	·	·	·	·	·

- na = no information supplied by the National Center.
- nc = not computed because of small Ns, but the standard errors are high.
- · = not tested and, therefore, no information required.

Appendix D

Test Scores for Tests 3E, 3X, and 3N (Population 3)

Test 3E Earth Science + Core Items (34 Items) Percent Score

Country	No. of students	Mean	Standard deviation	Standard error
Australia	181	65.5	15.4	1.66
Finland	891	55.9	12.6	.59
Hungary	123	68.1	12.6	1.51
Italy	737	61.7	13.9	1.03
Japan	464	63.7	13.1	1.65
Korea	1,054	60.3	15.2	.98

Test 3X General Science Test Only (30 Items) Percent Score

Country	No. of students	Mean	Standard deviation	Standard error
Italy	1,526	26.4	12.7	1.52
Papua New Guinea	487	30.6	8.5	.92
Singapore	345	40.9	11.1	2.33

Test 3N Scientific Understanding Only (30 Items) Percent Score

Country	No. of students	Mean	Standard deviation	Standard error
Australia	995	60.1	14.9	.66
Canada (Eng.)	509	56.8	14.9	1.01
Canada (Fr.)	889	51.0	13.5	.66
England	1,004	65.2	13.9	.58
Hungary	1,036	67.5	12.8	.93
Israel	388	49.1	17.3	2.59
Italy	2,455	58.9	15.0	.80
Japan	2,230	66.3	14.5	.94
Norway	595	64.7	13.8	.76
Papua New Guinea	291	51.3	14.2	1.43
Singapore	297	56.2	13.0	1.49
Sweden	1,281	67.1	13.1	.53
Thailand	3,605	50.3	12.4	.51

Appendix E Test Reliabilities and Standard Sampling Errors (in percentage points)

Country	Population 1 Core Test (24 items) KR-20	SE	Population 2 Core Test (30 items) KR-20	SE	Core Test (30 items) KR-20	SE	Population 3 3B (39 items) KR-20	SE	3C (39 items) KR-20	SE	3P (38 items) KR-20	SE	3N (30 items) KR-20	3X KR-20
Australia	.761	.73	.774	.63	.831	.56	.630a	.50	.801	.81	.707	.68	.717	-
Canada (Eng.)	.736	.54	.754	.56	.828	.48	.656	.63	.722	.62	.664b	.67	.712	-
Canada (Fr.)	.731	.67	.683	.67	.761	.64	.582	1.14	.517	.60	.418b	.57	.612	-
China	-	-	.754	.90	-	-	-	-	-	-	-	-	-	-
England	.755	.70	.771	.73	.833	.54	.624	.74	.798	.82	.713	.60	-	-
Finland	.713	.62	.703	.43	.790	.47	.780	.51	.832	.87	.850	.90	.599	-
Ghana	-	-	.745	1.20	.773	1.66	.624	1.20	.787	2.07	.602	1.43	-	-
Hong Kong	.716	.81	.716	.83	-	-	-	-	-	-	-	-	-	-
Hong Kong (Form 6)	-	-	-	-	.642	.66	.686	.83	.779	1.20	.706	1.08	-	-
Hong Kong (Form 7)	-	-	-	-	.610	.69	.747	.98	.841	1.15	.724	1.09	-	-
Hungary	.776	.96	.796	.85	.801	1.06	.663	1.13	.818	2.19	.805	1.50	-	-
Israel	.774	1.00	.763	1.14	.842	1.20	.881	1.98	.892	4.13	.818	1.42	.665	-
Italy	.793	1.06	-	-	.817	.81	.663	3.49	.890	4.76	.673	.89	.901	.698
Italy (Grade 8)	-	-	.754	.71	-	-	-	-	-	-	-	-	.711	-
Italy (Grade 9)	-	-	.763	1.43	-	-	-	-	-	-	-	-	-	-
Japan	.715	.28	.800	.29	.867	1.13	.690	1.70	.873	2.66	.785	1.85	-	-
Korea	.738	.66	.745	.48	.836	.78	.652	.56	.648	.56	.759	1.10	.651	-
Netherlands	-	-	.794	.86	-	-	-	-	-	-	-	-	-	-
Nigeria	.719	1.65	.753	1.08	-	-	-	-	-	-	-	-	-	-
Norway	.712	1.25	.677	.52	.837	.76	.693	.96	.769	1.16	.731	1.06	-	-
Papua New Guinea	-	-	.595	.57	.675	1.93	-	-	-	-	-	-	.644	.284
Philippines	.771	.66	.735	.67	-	-	-	-	-	-	-	-	.537	-
Poland	.768	.66	.794	.74	.779	.70	.591	.90	.780	1.34	.777	1.33	-	-
Singapore	.737	.74	.765	.93	.783	1.79	.630	1.44	.794	3.15	.637	1.31	-	.584
Sweden	-	-	-	-	.752	.83	.722	.77	.761	.70	.694	.63	.588	-
Sweden (Grade 3)	.739	.89	.754	.76	-	-	-	-	-	-	-	-	.617	-
Sweden (Grade 4)	.703	.65	.782	.74	-	-	-	-	-	-	-	-	-	-
Thailand	-	-	.664	.74	.822	.81	.607	.83	.729	1.17	.779	1.31	-	-
U.S.A.c	.770	.75	.770	.91	-	-	-	-	-	-	-	-	-	-
Zimbabwe	-	-	.738	.62	-	-	.669	1.68	.765	2.60	.695	2.07	.546	-

a Australia had only 38 items in the biology test because one item had a typing error in it which reduced its validity.
b Canada (Eng.) and Canada (Fr.) requested that one item be dropped from the physics test because one phrase had been omitted in the printed version of the item.
c In the United States 5 items were dropped from both the biology and chemistry tests and 4 items from the physics test. They also did not take the core test.

Appendix F1   Mean Scores, Standard Deviations and Standard Errors of Selected Subscores (in Percent): Biology, Chemistry, Earth Science, Physics (Population 1)*

	N	Biology (N=22)			Chemistry (N=5)			Earth (N=8)			Physics (N=21)		
		M	sd	se	M	sd	se	M	sd	se	M	sd	se
Australia	4259	58.3	19.0	.67	49.4	31.3	.87	58.9	22.2	.77	56.0	21.6	.77
Canada (Eng.)	5032	63.0	17.3	.46	58.1	31.1	.69	62.5	21.7	.61	60.5	19.9	.55
Canada (Fr.)	2676	62.8	16.8	.56	60.4	30.7	.95	60.9	22.0	.73	60.5	19.8	.70
England	3698	55.5	18.8	.63	46.7	31.6	.83	56.8	22.9	.77	52.1	20.9	.70
Finland	1600	67.9	16.8	.62	68.7	29.5	.99	64.6	21.0	.75	64.1	18.8	.57
Hong Kong	5342	55.9	17.3	.64	47.5	31.3	.97	50.9	22.4	.85	46.4	20.0	.81
Hungary	2590	64.5	18.4	.86	68.6	28.2	1.18	66.1	22.4	.99	55.3	20.3	.88
Israel	2351	43.5	17.5	.81	38.0	29.6	1.07	45.9	19.9	.98	37.7	17.2	.81
Italy	5152	62.5	19.2	.95	69.1	27.7	.97	63.6	24.3	1.40	51.3	20.8	1.17
Japan	7924	68.5	18.0	.27	51.4	31.5	.46	67.2	20.1	.31	67.5	17.4	.26
Korea	3489	65.9	18.2	.60	62.3	30.7	.83	63.7	22.0	.69	67.0	19.4	.67
Nigeria	944	35.8	19.0	1.66	35.1	29.2	2.26	36.4	21.4	1.73	33.5	18.4	1.55
Norway	1305	53.9	16.9	.77	62.7	31.7	1.91	57.2	22.5	1.63	54.6	21.1	1.44
Philippines	16851	44.1	18.7	.56	38.2	29.7	.72	46.1	22.9	.66	39.2	20.5	.67
Poland	4390	57.9	18.6	.60	55.1	31.5	.86	52.0	23.3	.78	46.2	21.1	.68
Singapore	5547	53.9	18.2	.74	40.3	30.4	.89	52.4	22.0	.76	51.6	19.5	.77
Sweden (Grade 3)	1327	59.0	17.2	.57	56.0	31.3	1.18	60.5	23.2	1.09	50.5	20.1	.93
Sweden (Grade 4)	1422	64.6	16.1	.79	63.7	26.6	.96	67.2	21.3	.93	58.9	19.3	.70

N = number of items   NOTE: *Since the United States did not administer the rotated tests, it was impossible to calculate the scores for the United States.

Appendix F2   Mean Scores, Standard Deviations and Standard Errors of Cognitive Behavioral Scores (in Percent): Information, Comprehension, Application (Population 1)*

	N	Information (N=20)			Comprehension (N=18)			Application (N=18)		
		M	sd	se	M	sd	se	M	sd	se
Australia	4259	60.1	18.3	.64	58.7	20.8	.74	52.7	21.5	.76
Canada (Eng.)	5032	65.0	17.1	.47	62.6	19.0	.51	58.1	20.2	.57
Canada (Fr.)	2676	62.4	17.0	.56	60.0	18.4	.66	62.3	20.1	.71
England	3698	58.9	17.7	.56	54.0	21.0	.72	49.5	21.0	.73
Finland	1600	67.9	16.6	.60	67.3	18.2	.59	63.2	19.2	.65
Hong Kong	5342	54.7	16.7	.59	50.0	19.8	.79	48.6	20.3	.85
Hungary	2590	60.6	17.1	.83	63.2	20.5	.95	61.6	20.3	.92
Israel	2351	35.5	16.7	.70	45.8	18.6	.97	42.5	17.5	.82
Italy	5152	63.9	18.2	.80	58.1	21.2	1.17	56.0	21.5	1.14
Japan	7924	60.1	16.6	.26	71.0	17.9	.28	67.9	19.2	.30
Korea	3489	64.5	16.6	.56	69.0	19.2	.64	63.9	20.8	.69
Nigeria	944	38.2	17.3	1.41	36.6	20.6	1.84	31.4	18.4	1.60
Norway	1305	62.1	18.3	1.02	54.2	19.4	1.07	51.0	19.7	1.38
Philippines	16851	49.3	18.3	.54	43.7	20.6	.66	35.2	20.3	.65
Poland	4390	52.6	18.9	.61	52.1	21.0	.69	53.0	20.6	.66
Singapore	5547	54.2	17.4	.68	55.2	20.4	.83	47.0	19.0	.74
Sweden (Grade 3)	1327	59.2	17.4	.70	57.3	19.7	.85	52.5	20.1	10
Sweden (Grade 4)	1422	63.7	16.2	.68	64.9	17.8	.62	60.6	19.4	.77

N = number of items   NOTE: *Since the United States did not administer the rotated tests, it was impossible to calculate the scores for the United States.

Appendix F3
Means, Standard Deviations and Standard Errors of Sampling of Science Subscores (in Percent) (Population 2)*

		Biology (N=23)			Chemistry (N=15)			Earth (N=9)			Physics (N=23)		
	N	M	sd	se	M	sd	se	M	sd	se	M	sd	se
Australia	4917	58.7	19.6	.76	50.1	24.0	.78	65.1	19.9	.56	60.9	16.8	.58
Canada (Eng.)	5382	63.5	17.8	.63	54.6	23.8	.87	65.2	19.3	.56	62.1	16.6	.54
Canada (Fr.)	2269	60.9	16.7	.79	49.2	21.8	.83	61.0	19.2	.75	60.2	15.9	.71
China	2806	52.5	16.5	.84	63.6	25.0	1.19	60.2	18.6	.84	65.5	17.6	.89
England	3069	54.9	19.2	.79	49.2	23.5	1.03	58.9	20.1	.64	59.5	17.2	.67
Finland	2546	60.8	16.4	.52	55.0	22.5	.62	64.5	19.3	.53	61.1	16.2	.46
Ghana	2766	47.9	18.2	1.20	44.5	23.0	1.44	48.3	20.0	1.15	46.3	18.3	1.25
Hong Kong	4973	52.9	17.5	.88	46.9	20.8	.92	63.8	19.4	.84	57.4	17.5	.98
Hungary	2515	69.1	17.0	.86	68.0	23.6	1.10	76.2	16.3	.65	71.9	18.7	.93
Israel	2082	58.7	19.1	1.23	56.8	26.3	1.75	66.6	19.8	.98	56.0	19.7	1.35
Italy (Grade 8)	4622	54.2	17.5	.60	43.6	21.6	.81	58.3	21.0	.65	52.7	19.0	.76
Italy (Grade 9)	1398	61.5	17.3	1.42	52.1	22.5	1.85	65.5	19.5	1.38	60.0	18.1	1.46
Japan	7610	65.0	19.0	.32	58.8	23.9	.38	68.9	20.1	.32	72.7	16.9	.26
Korea	4427	59.8	17.5	.49	53.3	21.9	.60	65.0	20.7	.58	64.8	17.6	.54
Netherlands	5025	60.0	17.7	.71	61.0	26.1	1.32	67.6	19.5	.72	67.7	18.7	.87
Nigeria	804	46.4	16.2	1.10	37.1	22.3	1.67	40.6	18.5	1.16	41.7	16.7	1.29
Norway	1420	58.1	17.9	.65	55.4	21.5	.72	63.2	21.3	.84	61.3	17.1	.55
Papua New Guinea	2115	56.8	14.8	.62	45.7	18.7	.78	59.7	18.4	.73	56.9	14.5	.57
Philippines	10871	41.4	18.0	.71	35.5	19.8	.68	43.4	19.8	.67	38.5	17.1	.68
Poland	4517	61.1	17.7	.65	59.6	24.7	.93	63.7	19.7	.64	56.4	19.0	.74
Singapore	4430	55.8	19.4	1.04	49.7	24.7	1.33	59.6	19.5	.80	59.2	16.9	.87
Sweden (Grade 7)	1493	54.0	17.9	.69	50.5	22.2	1.00	64.2	19.8	.78	57.1	17.7	.87
Sweden (Grade 8)	1363	58.5	18.8	.86	56.8	24.4	1.13	66.0	20.0	.77	61.7	18.0	.91
Thailand	3780	60.0	14.3	.67	47.8	20.0	.94	59.0	20.0	.82	57.2	15.5	.78
Zimbabwe	2648	45.4	15.4	.77	36.8	18.4	.97	45.6	19.0	.68	42.1	14.4	.70

N = number of items

NOTE: *Since the United States did not administer the rotated tests, it was impossible to calculate the scores for the United States.

Appendix F4
Means, Standard Deviations and Standard Errors of Sampling of Cognitive Behavioral Scores (in Percent) (Population 2)*

		Information (N=20)			Comprehension (N=18)			Application (N=18)		
	N	M	sd	se	M	sd	se	M	sd	se
Australia	4917	57.2	19.2	.65	58.8	17.4	.59	59.7	18.6	.70
Canada (Eng.)	5383	60.9	18.3	.59	61.0	16.9	.61	62.3	17.2	.63
Canada (Fr.)	2269	54.8	17.0	.66	58.8	16.2	.72	60.5	16.1	.81
China	2806	56.3	17.7	.86	67.0	18.7	.96	57.1	16.7	.87
England	3069	52.4	18.4	.65	56.7	17.4	.70	57.4	18.7	.84
Finland	2546	57.9	17.3	.49	61.6	16.0	.46	60.8	16.5	.54
Ghana	2766	48.3	19.4	1.23	48.8	19.1	1.29	44.2	16.6	1.14
Hong Kong	4973	50.4	16.4	.74	55.4	17.6	.96	57.4	17.0	.93
Hungary	2515	71.4	16.9	.81	67.9	19.1	.89	72.2	16.6	.87
Israel	2062	55.9	19.3	1.15	59.9	20.3	1.43	59.1	18.4	1.28
Italy (Grade 8)	4622	52.5	18.1	.65	52.4	18.4	.70	52.0	17.6	.71
Italy (Grade 9)	1398	57.1	17.9	1.40	61.5	17.4	1.52	60.0	17.3	1.45
Japan	7610	59.3	19.4	.33	69.6	17.9	.28	69.4	18.0	.29
Korea	4427	59.7	18.1	.54	61.8	18.2	.51	61.2	17.1	.51
Netherlands	5025	60.8	18.4	.76	65.8	19.3	.92	64.0	18.0	.86
Nigeria	804	44.0	17.8	1.36	44.2	18.0	1.31	39.8	15.1	1.11
Norway	1420	60.3	16.3	.52	59.1	17.2	.56	58.9	17.8	.61
Papua New Guinea	2115	55.5	15.7	.66	56.5	14.6	.58	54.3	13.7	.62
Philippines	10871	44.2	18.7	.70	43.2	17.6	.68	34.3	15.4	.62
Poland	4517	59.7	19.4	.73	60.7	19.3	.73	58.6	17.1	.67
Singapore	4430	56.2	17.8	.87	59.8	17.8	.93	53.9	19.1	1.07
Sweden (Grade 7)	1493	51.1	18.3	.86	56.6	17.1	.64	58.2	17.4	.76
Sweden (Grade 8)	1363	55.8	19.0	.88	60.5	17.9	.85	62.7	18.1	.88
Thailand	3780	53.9	15.0	.62	60.6	16.2	.83	55.5	15.6	.81
Zimbabwe	2648	41.9	15.6	.77	45.3	15.3	.71	41.3	14.2	.67

N = number of items

NOTE: *Since the United States did not administer the rotated tests, it was impossible to calculate the scores for the United States.

Appendix G   Report on Attitude and Descriptive Scales[1]

Introduction

Where a country had not administered the items or had too few items (e.g. United States Phase 2) then the country was not included in the analyses.

The following steps were undertaken:
1. A principal components analysis was undertaken for each set of items said to belong to each scale for each country at each population level.
2. Where a factor loading for an item was below .3 then the item was examined in terms of whether it could belong to the scale or not. In general, the item was dropped from the scale for all countries where unfactorloading fell below 0.3 for more than two countries.
3. Even though the reliability of a scale was low at the between student level, the scale was nevertheless retained on the assumption that at the aggregated level of between schools or classrooms the reliability would be higher.

Part 1 of this appendix presents for each population level the titles of the scales and a note in parentheses on the countries for which scale scores were not calculated for that country either becausea country had not administered the scale or because an item had been dropped. Part 2 of the appendix presents the means, standard deviations and alpha coefficients at the between student level, as well as the N used for each scale in each country at each population level.

Part 1:   Items and Scales

Notes
ATT   attitude scale
DES   descriptive scale
(r)   the direction of the scoring of the item was reversed.
The scale range in the international version of the measure is also given.

Population I

Scale 1, STUDENT PARTICIPATION (Range: 3-9)
P1DES02 We use library books for learning science. *(Philippines)*
P1DES03 We choose the topics we want to study in science.
P1DES04 The teacher uses our ideas and suggestions when planning science lessons.

Scale 2, PRACTICAL WORK (Range: 3-12)
P1DES05 We watch the teacher do experiments during our science lessons. *(Japan, Korea and Singapore)*
P1DES09 The students themselves do experiments as part of the science lessons.
P1DES10 The class breaks into small groups of students to do experiments during science lessons.

---

[1] This appendix was prepared by Christian Morgenstern of Hamburg University.

Scale 3, IMPORTANCE OF SCIENCE (Range: 6-18)
(Israel not included)
P1ATT03 Scientific inventions improve our standard of living.
(r)P1ATT05 Science has ruined the environment. *(Sweden Grades 3 and 4)*
P1ATT10 Science will help to make the world a better place in the future.
(r)P1ATT12 Scientific discoveries do more harm than good.
P1ATT15 Science is very important for a country's development.
P1ATT17 The government should spend more money on scientific research.

Scale 4, INTEREST IN SCIENCE (Range: 4-12)
(Israel not included)
LIKESCI I like science.
P1ATT01 Science is an enjoyable school subject.
P1ATT06 The science taught at school is interesting.
P1DES06 The teacher makes science lessons interesting for us.

Scale 5, LIKE SCHOOL (Range: 7-21)
(Israel and Sweden not included)
(r)P1ATT04 I generally dislike my schoolwork. *(Norway)*
(r)P1ATT07 School is not very enjoyable.
P1ATT09 I want as much education as I can get. *(Japan)*
P1ATT11 I enjoy everything about school.
(r)P1ATT13 I am bored most of the time in school.
(r)P1ATT16 There are many school subjects I don't like.
P1ATT18 The most enjoyable part of my life is the time I spend at school.

Scale 6, FACILITY LEARNING SCIENCE (Range: 2-6)
(Nigeria, Philippines and Sweden Grades 3 and 4 not included)
(r)P1ATT08 Science is a difficult subject.
(r)P1ATT14 There are too many facts to learn in science.

Population 2

Scale 1, BENEFICIAL ASPECTS OF SCIENCE (Range: 7-21)
(Norway and Zimbabwe not included)
P2ATT01 Science is useful for solving the problems of everyday life.
P2ATT05 Science is very important for a country's development.
P2ATT07 Money spent on science is well worth spending.
P2ATT12 Public money spent on science in the last few years has been used wisely. *(China)*
P2ATT15 Scientific inventions improve our standard of living.
P2ATT16 The government should spend more money on scientific research. *(China, Hong Kong)*
P2ATT22 Science will help to make the world a better place in the future.

Scale 2, INTEREST IN SCIENCE (Range: 4-12)
(Israel, Italy Grades 8 and 9, and Korea not included)
LIKESCI I like science. *(China)*
P2ATT34 Science is an enjoyable school subject.
P2ATT35 The science taught at school is interesting.
P2DES09 The teacher makes science lessons interesting.

Scale 3, LIKE SCHOOL (Range: 7-21)
(Norway and Zimbabwe not included)
(r)P2ATT06 School is not very enjoyable.
P2ATT09 I enjoy everything about school. *(Ghana, Nigeria)*
(r)P2ATT14 I am bored most of the time at school.
(r)P2ATT19 There are many school subjects I don't like.
P2ATT23 The most enjoyable part of my life is the time I spend at school.

P2ATT26 I want as much education as I can get.
(r)P2ATT28 I generally dislike my schoolwork.

Scale 4, NON-HARMFUL ASPECTS OF SCIENCE (Range: 6-18)
(Israel and Zimbabwe not included)

(r)P2ATT03 Science has ruined the environment.
(r)P2ATT10 Much of the anxiety in modern society is due to science.
(r)P2ATT17 Scientific inventions have made the world too complex.
(r)P2ATT20 Scientific inventions have increased tensions between people.
(r)P2ATT24 Scientific discoveries do more harm than good.
(r)P2ATT27 Science and technology are the cause of many of the world's problems.

Scale 5, CAREER INTEREST IN SCIENCE (Range: 7-21)
(Israel and Zimbabwe not included)

P2ATT04 Working in a science laboratory would be an interesting way to earn a living.
P2ATT08 In the future most jobs will require a knowledge of science.
P2ATT11 People who understand science are better off in our society.
P2ATT13 It is important to know science in order to get a good job.
P2ATT18 Science is a very good field for creative people to enter.
P2ATT21 In my future career I would like to use the science I learned at school.
P2ATT29 I want to learn more about world we live in.

Scale 6, FACILITY OF LEARNING SCIENCE (Range: 4-12)

(r)P2ATT36 Science is a difficult subject.
(r)P2ATT37 Science is difficult when it involves calculations. *(Norway)*
(r)P2ATT38 Science is difficult when it involves handling apparatus. *(Korea, Zimbabwe)*
(r)P2ATT39 There are too many facts to learn in science. *(Nigeria, Papua New Guinea, Philippines)*

Scale 7, TEACHER DIRECTED LEARNING (Range: 6-18)
(Korea and Zimbabwe not included)

P2DES03 At the start of each science lesson the teacher reminds us about the work we covered during previous lessons.
P2DES04 At the start of each science lesson the teacher explains the work we have to cover during the lesson.
P2DES05 At the end of each science lesson the teacher gives a summary of what was taught in the lesson.
P2DES08 The teacher does demonstrations to help explain scientific ideas.
P2DES12 The teacher explains how the science we learn is relevant to our own lives.
P2DES15 The science teacher helps students who have difficulties with learning science.

Scale 8, STUDENT PARTICIPATION (Range: 6-18)

P2DES06 We are allowed to make our own choice of science topics to study.
P2DES07 The teacher uses our ideas and suggestions when planning science lessons.
P2DES17 We do field work outside the classroom as part of our science lessons.
P2DES22 In our practical work we make up our own problems and then the teacher helps us to plan experiments to solve them. *(Zimbabwe)*
P2DES23 When we do experiments the teacher gives us problems to solve and then leaves us to work out our own methods and solutions. *(China)*
P2DES24 In our practical work we make up our own problems and work out our own methods to investigate the problems.

Scale 9, PRACTICAL WORK (Range: 5-15)
(Sweden Grade 8 and Zimbabwe not included)

P2DES11 For science homework we write up reports of our laboratory and practical work. *(Sweden Grade 7)*
P2DES18 We do practical work (experiments) as part of our science lessons.
P2DES19 The science class breaks up into small groups of students to do practical work (experiments).
P2DES20 When we do experiments the teacher gives us instructions about what to do.

P2DES21   When we do an experiment we use a book or other written instructions to show us how to do it.

Scale 10, HOMEWORK EFFORT (Range: 0-18)
(Australia, Canada (Eng.), Canada (Fr.), China, England, Finland, Japan, Korea, Norway, Papua New Guinea, Zimbabwe not included)
P2DES32   When you have finished an assignment, do you check it?
P2DES33   When you have an assignment, do you do your best?
P2DES35   When you sit down to do your homework, how long does it take you to get started? *(Hong Kong)*
P2DES36   When you do your homework, are you disturbed by things going on around you?
P2DES37   When you have homework, do you do all of it?
P2DES38   When you have homework, do you hand it in on time?

Scale 11, CLASSROOM EFFORT (Range: 0-15)
(Australia, Canada (Eng.), Canada (Fr.), England, Finland, Hong Kong, Korea, Norway, Papua New Guinea and Zimbabwe not included)
P2DES25   What do you do when the teacher asks you a question in class?
P2DES26   What do you do when the teacher asks someone else a question?
P2DES27   When the teacher is preparing something new, do you pay attention?
P2DES39   When you are absent and miss a lesson, do you try to make up what you have missed?
P2DES40   Do you ever go to school but deliberately miss a lesson? *(China)*

Population 3

Scale 1, BENEFICIAL ASPECTS OF SCIENCE (Range: 7-21)
(Israel and Norway not included)
P3ATT01   Science is useful for solving the problems of everyday life.
P3ATT05   Science is very important for a country's development.
P3ATT07   Money spent on science is well worth spending.
P3ATT12   Public money spent on science in the last few years has been used wisely.
P3ATT15   Scientific inventions improve our standard of living.
P3ATT16   The government should spend more money on scientific research. *(Hong Kong Forms 6 and 7)*
P3ATT22   Science will help to make the world a better place in the future.

Scale 2, INTEREST IN SCIENCE (Range: 8-24)
(Canada (Fr.), Israel and Norway not included)
LIKESCI   I like science.
P3ATT36   Biology is an enjoyable school subject.
P3ATT37   Chemistry is an enjoyable school subject.
P3ATT38   Physics is an enjoyable school subject.
P3ATT39   The biology taught at school is interesting.
P3ATT40   The chemistry taught at school is interesting.
P3ATT41   The physics taught at school is interesting.
P3DES09   The teacher makes science lessons interesting for us. *(Italy)*

Scale 3, LIKE SCHOOL (Range: 7-21)
(Israel, Norway and Papua New Guinea not included)
(r)P3ATT06   School is not very enjoyable.
P3ATT09   I enjoy everything about school.
(r)P3ATT14   I am bored most of the time at school.
(r)P3ATT19   There are many school subjects I don't like.
P3ATT23   The most enjoyable part of my life is the time I spend at school.
P3ATT26   I want as much education as I can get.
(r)P3ATT28   I generally dislike my schoolwork.

Scale 4, NON-HARMFUL ASPECTS OF SCIENCE (Range: 6-18)
(Israel and Norway not included)

(r)P3ATT03 Science has ruined the environment.
(r)P3ATT10 Much of the anxiety in modern society is due to science.
(r)P3ATT17 Scientific inventions have made the world too complex.
(r)P3ATT20 Scientific inventions have increased tensions between people.
(r)P3ATT24 Scientific discoveries do more harm than good.
(r)P3ATT27 Science and technology are the cause of many of the world's problems.

Scale 5, CAREER INTEREST IN SCIENCE (Range: 7-21)
(Israel and Norway not included)

P3ATT04 Working in a science laboratory would be an interesting way to earn a living.
P3ATT08 In the future most jobs will require a knowledge of science.
P3ATT11 People who understand science are better off in our society.
P3ATT13 It is important to know science in order to get a good job.
P3ATT18 Science is a very good field for creative people to enter.
P3ATT21 In my future career I would like to use the science I learned at school.
P3ATT25 People who work with modern inventions such as computers have more interesting jobs.

Scale 6, FACILITY LEARNING SCIENCE (Range: 12-36)
(Canada (Fr.), Hungary, Israel and Norway not included)

(r)P3ATT42 Biology is a difficult subject.
(r)P3ATT43 Chemistry is a difficult subject.
(r)P3ATT44 Physics is a difficult subject.
(r)P3ATT45 Biology is difficult when it involves calculations.
(r)P3ATT46 Chemistry is difficult when it involves calculations.
(r)P3ATT47 Physics is difficult when it involves calculations.
(r)P3ATT48 Biology is difficult when it involves handling apparatus.
(r)P3ATT49 Chemistry is difficult when it involves handling apparatus.
(r)P3ATT50 Physics is difficult when it involves handling apparatus.
(r)P3ATT51 There are too many facts to learn in biology.
(r)P3ATT52 There are too many facts to learn in chemistry.
(r)P3ATT53 There are too many facts to learn in physics.

Scale 7, TEACHER DIRECTED LEARNING (Range: 6-18)
(Israel and Norway not included)

P3DES03 At the start of each science lesson the teacher reminds us about the work we covered during previous lessons.
P3DES04 At the start of each science lesson the teacher explains the work we have to cover during the lesson.
P3DES05 At the end of each science lesson the teacher gives a summary of what was taught in the lesson.
P3DES08 The teacher does demonstrations to help explain scientific ideas.
P3DES12 The teacher explains how the science we learn is relevant to our own lives.
P3DES15 The science teacher helps students who have difficulties with learning science.

Scale 8, STUDENT PARTICIPATION (Range: 6-18)
(Israel and Norway not included)

P3DES06 We are allowed to make our own choice of science topics to study.
P3DES07 The teacher uses our ideas and suggestions when planning science lessons.
P3DES17 We do field work outside the classroom as part of our science lessons.
P3DES22 In our practical work we make up our own problems and then the teacher helps us to plan experiments to solve them.
P3DES23 When we do experiments the teacher gives us problems to solve and then leaves us to work out our own methods and solutions.
P3DES24 In our practical work we make up our own problems and work out our own methods to investigate the problems.

Scale 9, PRACTICAL WORK (Range: 5-15)
(Israel and Norway not included)

- P3DES11 For science homework we write up reports of our laboratory and practical work.
- P3DES18 We do practical work (experiments) as part of our science lessons.
- P3DES19 The science class breaks up into small groups of students to do practical work (experiments).
- P3DES20 When we do experiments the teacher gives us instructions about what to do.
- P3DES21 When we do an experiment we use a book or other written instructions to show us how to do it.

Scale 10, HOMEWORK EFFORT (Range: 0-18)
(Australia, Canada (Eng.), Canada (Fr.), England, Ghana, Hong Kong (Forms 6 and 7), Israel, Japan, Norway, Papua New Guinea and Sweden not included)

- P3DES32 When you have finished an assignment, do you check it?
- P3DES33 When you have an assignment, do you do your best?
- P3DES35 When you sit down to do your homework, how long does it take you to get started?
- P3DES36 When you do your homework, are you disturbed by things going on around you?
- P3DES37 When you have homework, do you do all of it?
- P3DES38 When you have homework, do you hand it in on time?

Scale 11, CLASSROOM EFFORT (Range: 0-15)
(Australia, Canada (Eng.), Canada (Fr.), England, Ghana, Israel, Norway, Papua New Guinea and Sweden not included)

- P3DES25 What do you do when the teacher asks you a question in class?
- P3DES26 What do you do when the teacher asks someone else a question?
- P3DES27 When the teacher is preparing something new, do you pay attention?
- P3DES39 When you are absent and miss a lesson, do you try to make up what you have missed?
- P3DES40 Do you ever go to school but deliberately miss a lesson?

Part 2: **Means, Standard Deviations, N, and Alpha Coefficients for Each Scale in Each Country at Each Population Level**

Computation of scores

Three response categories were, in general, used: 'agree,' 'disagree' and 'uncertain.' The scoring scheme employed involved the scoring of a favourable response to an item- 3; an uncertain response- 2; and an unfavourable response- 1. At the Population 3 level some countries used five response categories which have been reduced to three categories to maintain consistency with other countries. The scoring of an agree or disagree response was reversed for items marked (r). High values indicate a positive attitude whereas low values represent a negative attitude. Special recoding was applied for selected descriptive items in Populations 2 and 3, namely: P3DES25 through P3DES40. Starting with item P3DES25 the scale range was collapsed to four points (0,1,2,3) from the original five points. As a result of such collapsing, an equal item and scale range could be produced. Following the data preparation an extensive data check was performed to check the valid item range after recoding and on item availability across countries. Where a country failed to administer a particular item in a particular scale, scale scores for that scale were not calculated for that country. The same procedure was used in cases where countries had items with high missing values. After these steps had been accomplished, standardized scale values were computed by dividing each national scale by the number of items in it. The concluding section presents the mean, sd, N of students, and alpha coefficient for each scale in each country. The alpha values have been calculated at the between-student level. Even though the alpha values are, in some cases, very low some scales may be used at the between school level analysis although not at the between student level.

Population 1

Country	Student Participation				Practical Work				Importance of Science			
	M	sd	N	α	M	sd	N	α	M	sd	N	α
Australia	1.75	.45	4164	.42	1.94	.52	4160	.54	2.44	.39	4136	.55
Canada (Eng.)	1.63	.41	5065	.42	2.00	.49	5057	.53	2.50	.36	5051	.58
Canada (Fr.)	1.56	.45	2619	.48	1.63	.52	2620	.59	2.50	.39	2655	.58
England	1.75	.45	3623	.30	2.06	.51	3611	.48	2.38	.41	3571	.58
Finland	1.69	.38	1560	.19	1.69	.45	1559	.32	2.38	.35	1586	.53
Hong Kong	2.06	.37	3388	.49	2.13	.40	3256	.62	1.69	.25	5279	.22
Hungary	1.75	.37	2579	.24	1.94	.43	2577	.42	2.63	.32	2553	.52
Israel	1.44	.40	2130	.39	2.25	.45	2125	.40	n.c.	n.c.	n.c.	n.c.
Italy	1.75	.43	5072	.34	1.75	.50	5058	.51	2.38	.37	5050	.55
Japan	1.56	.41	6602	.41	n.c.	n.c.	n.c.	n.c.	2.25	.37	6600	.52
Korea	1.69	.42	3484	.35	n.c.	n.c.	n.c.	n.c.	2.56	.30	3478	.44
Nigeria	1.81	.49	905	.26	2.06	.51	890	.44	2.44	.44	888	.59
Norway	1.75	.42	1237	.42	1.94	.50	1214	.51	2.44	.36	1223	.47
Philippines	n.c.	n.c.	n.c.	n.c.	2.13	.43	6795	.30	2.25	.41	16710	.54
Poland	1.81	.45	4364	.48	1.94	.43	4362	.45	2.44	.32	4353	.49
Singapore	1.63	.41	5531	.24	n.c.	n.c.	n.c.	n.c.	2.56	.34	5515	.50
Sweden (Grade 3)	1.94	.34	1215	.35	1.69	.46	1187	.59	n.c.	n.c.	n.c.	n.c.
Sweden (Grade 4)	1.88	.36	1352	.34	1.69	.50	1342	.68	n.c.	n.c.	n.c.	n.c.

n.c. = not calculated because of missing data

Population 1 ctd.

Country	Interest in Science				Like School				Facility of Learning Science			
	M	sd	N	α	M	sd	N	α	M	sd	N	α
Australia	2.38	.47	4165	.63	2.19	.52	4134	.74	2.19	.68	4168	.40
Canada (Eng.)	2.44	.50	5065	.72	2.38	.48	5051	.76	2.19	.70	5566	.50
Canada (Fr.)	2.50	.43	2653	.66	2.19	.52	2656	.75	2.50	.63	2654	.47
England	2.38	.49	3623	.63	2.19	.53	3584	.75	2.06	.70	3629	.44
Finland	2.44	.45	1590	.66	2.00	.52	1585	.79	2.31	.61	1592	.35
Hong Kong	2.44	.38	5280	.51	2.44	.40	5250	.65	1.75	.57	5287	.17
Hungary	2.56	.33	2557	.47	2.31	.43	2553	.61	2.19	.74	2581	.46
Israel	n.c.	n.c.	n.c.	n.c.	n.c.	n.c.	n.c.	n.c.	1.75	.62	2151	.41
Italy	2.63	.32	5094	.53	2.31	.45	5067	.67	2.19	.67	5090	.44
Japan	2.50	.47	6611	.71	n.c.	n.c.	n.c.	n.c.	2.50	.58	6612	.35
Korea	2.44	.41	3488	.55	2.25	.42	3479	.58	2.00	.60	3487	.21
Nigeria	2.50	.42	908	.46	2.31	.42	891	.52	n.c.	n.c.	n.c.	n.c.
Norway	2.44	.46	1249	.75	n.c.	n.c.	n.c.	n.c.	2.31	.65	1245	.51
Philippines	2.56	.38	16743	.38	2.44	.42	16695	.62	n.c.	n.c.	n.c.	n.c.
Poland	2.56	.33	4364	.52	2.38	.41	4355	.63	2.00	.68	4368	.37
Singapore	2.56	.37	5539	.49	2.56	.38	5521	.64	1.94	.65	5541	.34
Sweden (Grade 3)	2.38	.44	1236	.68	n.c.	n.c.	n.c.	n.c.	n.c.	n.c.	n.c.	n.c.
Sweden (Grade 4)	2.31	.46	1366	.70	n.c.	n.c.	n.c.	n.c.	n.c.	n.c.	n.c.	n.c.

n.c. = not calculated because of missing data

Population 2

Country	Beneficial Aspects of Science				Interest in Science				Like School			
	M	sd	N	α	M	sd	N	α	M	sd	N	α
Australia	2.31	.40	4878	.62	2.13	.59	4862	.78	1.94	.52	4867	.74
Canada (Eng.)	2.44	.36	5437	.63	2.25	.58	5416	.80	2.00	.53	5409	.78
Canada (Fr.)	2.44	.36	2226	.64	2.25	.57	2220	.79	2.00	.55	2225	.79
China	n.c.	n.c.	n.c.	n.c.	n.c.	n.c.	n.c.	n.c.	2.44	.32	2709	.59
England	2.38	.39	2984	.64	2.19	.57	2970	.77	2.00	.52	2974	.77
Finland	2.31	.37	2500	.62	2.06	.57	2492	.76	1.63	.51	2498	.78
Ghana	1.88	.24	2617	.49	1.94	.34	2595	.43	n.c.	n.c.	n.c.	n.c.
Hong Kong	n.c.	n.c.	n.c.	n.c.	n.c.	n.c.	n.c.	n.c.	2.25	.43	4924	.66
Hungary	2.63	.26	2380	.44	2.25	.47	4939	.65	1.88	.44	2381	.65
Israel	2.50	.36	768	.64	2.50	.41	2401	.50	2.00	.43	719	.63
Italy (Grade 8)	2.50	.36	4484	.62	n.c.	n.c.	n.c.	n.c.	2.06	.50	4487	.71
Italy (Grade 9)	2.50	.36	1370	.63	n.c.	n.c.	n.c.	n.c.	2.13	.45	1371	.69
Japan	2.07	.39	6427	.60	1.88	.57	6430	.74	2.06	.46	6421	.67
Korea	2.69	.24	4514	.49	n.c.	n.c.	n.c.	n.c.	2.13	.46	4510	.62
Netherlands	2.18	.38	4801	.61	1.94	.53	4799	.69	1.81	.48	4803	.72
Nigeria	2.69	.29	749	.59	2.75	.29	731	.43	n.c.	n.c.	n.c.	n.c.
Norway	n.c.	n.c.	n.c.	n.c.	2.13	.61	1408	.80	n.c.	n.c.	n.c.	n.c.
Papua New Guinea	2.25	.34	1986	.50	n.c.	n.c.	n.c.	n.c.	2.56	.33	1992	.44
Philippines	2.50	.34	10823	.52	2.50	.35	1993	.51	2.56	.38	10809	.55
Poland	2.69	.34	4486	.46	2.56	.38	10840	.45	2.19	.46	4483	.68
Singapore	2.50	.34	4411	.55	2.19	.51	4373	.69	2.38	.44	4403	.66
Sweden (Grade 7)	2.25	.35	718	.52	2.19	.53	717	.69	1.94	.48	718	.73
Sweden (Grade 8)	2.25	.34	656	.52	2.06	.55	658	.74	1.81	.48	657	.72
Thailand	2.56	.29	3780	.42	2.38	.40	3779	.51	2.50	.36	3780	.49
Zimbabwe	n.c.	n.c.	n.c.	n.c.	2.44	.41	2084	.81	n.c.	n.c.	n.c.	n.c.

n.c. = not calculated because of missing data

Population 2 ctd.

Country	Non-harmful Aspects of Science				Career Interest in Science				Facility of Learning Science			
	M	sd	N	α	M	sd	N	α	M	sd	N	α
Australia	2.13	.43	4867	.59	2.13	.45	4874	.63	1.94	.54	4868	.51
Canada (Eng.)	2.19	.43	5408	.64	2.31	.41	5410	.61	2.13	.56	5405	.55
Canada (Fr.)	2.00	.45	2223	.64	2.38	.39	2227	.59	2.00	.56	2224	.58
China	2.19	.38	2706	.48	2.00	.16	2710	.40	1.50	.43	2711	.53
England	2.13	.44	2969	.62	2.25	.43	2977	.59	1.94	.52	2969	.49
Finland	2.19	.42	2503	.64	2.13	.39	2499	.57	2.13	.48	2486	.45
Ghana	1.94	.43	2604	.38	1.81	.27	2604	.48	1.81	.52	2584	.41
Hong Kong	1.94	.41	4922	.53	2.25	.34	4933	.49	1.75	.47	4929	.47
Hungary	2.25	.41	2380	.57	2.44	.30	2381	.46	1.88	.55	2400	.44
Israel	n.c.	n.c.	n.c.	n.c.	n.c.	n.c.	n.c.	n.c.	n.c.	n.c.	n.c.	n.c.
Italy (Grade 8)	2.13	.50	4474	.71	2.44	.35	4484	.51	2.13	.51	684	.44
Italy (Grade 9)	2.13	.51	1369	.74	2.44	.33	1368	.52	2.00	.53	4492	.48
Japan	2.00	.46	6418	.64	1.94	.45	6429	.63	1.94	.52	1360	.47
Korea	2.06	.50	4512	.63	2.38	.33	4513	.40	1.81	.56	6424	.52
Netherlands	2.13	.46	4796	.67	2.06	.45	4807	.63	n.c.	n.c.	n.c.	n.c.
Nigeria	2.00	.46	740	.58	2.69	.34	749	.63	1.88	.55	4805	.58
Norway	2.06	.46	1390	.65	1.50	.29	1392	.41	n.c.	n.c.	n.c.	n.c.
Papua New Guinea	1.94	.44	1990	.53	2.25	.36	1991	.50	n.c.	n.c.	n.c.	n.c.
Philippines	2.06	.41	10628	.43	2.56	.36	10831	.54	n.c.	n.c.	n.c.	n.c.
Poland	2.00	.43	4480	.61	2.56	.27	4482	.48	2.00	.53	4480	.54
Singapore	1.88	.38	4406	.42	2.25	.37	4402	.50	1.81	.49	4357	.42
Sweden (Grade 7)	1.88	.42	717	.58	2.25	.37	716	.54	1.94	.52	717	.51
Sweden (Grade 8)	1.88	.43	657	.58	2.19	.37	659	.49	1.94	.53	656	.54
Thailand	2.13	.44	3779	.50	2.63	.29	3779	.39	1.75	.50	3779	.42
Zimbabwe	n.c.	n.c.	n.c.	n.c.	n.c.	n.c.	n.c.	n.c.	n.c.	n.c.	n.c.	n.c.

n.c. = not calculated because of missing data

Report on Attitude and Descriptive Scales

Population 2 ctd.

Country	Teacher Directed Learning				Student Participation				Practical Work			
	M	sd	N	α	M	sd	N	α	M	sd	N	α
Australia	2.13	.39	4864	.62	1.44	.32	4856	.60	2.38	.36	4853	.54
Canada (Eng.)	2.25	.38	5369	.66	1.38	.30	5335	.57	2.44	.44	5358	.70
Canada (Fr.)	2.19	.38	2218	.63	1.50	.35	2212	.59	2.38	.43	2222	.69
China	2.56	.35	2701	.70	n.c.	n.c.	n.c.	n.c.	2.56	.37	2692	.66
England	2.19	.36	2976	.54	1.38	.33	2969	.61	2.44	.37	2972	.60
Finland	1.88	.36	2452	.57	1.50	.31	2446	.52	2.06	.45	2450	.68
Ghana	2.44	.32	2604	.45	1.44	.33	2596	.51	2.31	.39	2599	.53
Hong Kong	2.25	.36	4901	.61	1.75	.34	4896	.55	2.13	.34	4892	.76
Hungary	2.56	.32	2398	.59	1.75	.36	2396	.64	2.13	.34	2398	.58
Israel	2.25	.35	917	.53	1.56	.37	882	.64	2.31	.45	885	.68
Italy (Grade 8)	2.19	.30	4484	.19	1.94	.48	4401	.55	1.81	.51	4451	.61
Italy (Grade 9)	2.13	.35	1203	.30	1.69	.51	1135	.65	1.56	.55	1142	.72
Japan	1.94	.39	6418	.67	1.50	.33	4610	.60	2.38	.30	4622	.45
Korea	n.c.	n.c.	n.c.	n.c.	2.19	.40	4513	.54	1.81	.32	4512	.52
Netherlands	1.88	.37	4574	.54	1.25	.30	4539	.62	2.00	.59	4564	.79
Nigeria	2.50	.34	768	.62	1.88	.39	738	.57	2.25	.41	759	.61
Norway	2.06	.34	1408	.48	1.44	.34	1401	.58	2.31	.40	1408	.55
Papua New Guinea	2.50	.29	1986	.49	1.50	.32	1983	.55	2.38	.31	1987	.53
Philippines	2.44	.33	10758	.50	1.88	.36	10753	.46	2.50	.34	10770	.49
Poland	2.44	.35	4482	.64	1.75	.38	4479	.71	2.00	.41	4478	.65
Singapore	2.19	.38	3492	.61	1.44	.32	4360	.54	2.44	.36	4375	.54
Sweden (Grade 7)	2.13	.34	713	.48	1.56	.34	698	.62	n.c.	n.c.	n.c.	n.c.
Sweden (Grade 8)	2.13	.34	655	.51	1.50	.32	642	.60	n.c.	n.c.	n.c.	n.c.
Thailand	2.44	.31	3779	.54	1.81	.34	3787	.52	2.56	.30	3780	.48
Zimbabwe	n.c.	n.c.	n.c.	n.c.	n.c.	n.c.	n.c.	n.c.	n.c.	n.c.	n.c.	n.c.

n.c. = not calculated because of missing data

Appendix G

Population 2 ctd.

Country	Homework Effort				Classroom Effort			
	M	sd	N	α	M	sd	N	α
China	n.c.	n.c.	n.c.	n.c.	n.c.	n.c.	n.c.	n.c.
Ghana	1.75	.26	2521	.55	2.06	.33	2571	.46
Hong Kong	n.c.	n.c.	n.c.	n.c.	n.c.	n.c.	n.c.	n.c.
Hungary	2.44	.35	2392	.65	2.44	.33	2395	.59
Israel	2.25	.40	788	.61	2.31	.38	857	.44
Italy (Grade 8)	2.38	.46	4501	.76	2.38	.43	4305	.62
Italy (Grade 9)	2.44	.41	1287	.70	2.38	.40	1207	.63
Japan	n.c.	n.c.	n.c.	n.c.	2.13	.49	6401	.61
Korea	n.c.	n.c.	n.c.	n.c.	n.c.	n.c.	n.c.	n.c.
Netherlands	2.06	.54	4592	.77	2.13	.57	4543	.67
Nigeria	2.50	.41	720	.70	2.50	.43	718	.52
Philippines	2.38	.41	10760	.60	2.44	.41	10768	.43
Poland	2.44	.38	4470	.67	2.38	.38	4468	.58
Singapore	2.31	.38	4179	.60	2.31	.42	4171	.49
Sweden (Grade 7)	2.31	.53	693	.72	2.31	.48	693	.63
Sweden (Grade 8)	2.25	.58	648	.78	2.25	.53	647	.69
Thailand	2.31	.35	3774	.56	2.56	.34	3773	.42

n.c. = not calculated because of missing data

Population 3

Country	Beneficial Aspects of Science				Interest in Science				Like School			
	M	sd	N	α	M	sd	N	α	M	sd	N	α
Australia	2.44	.40	4958	.64	2.06	.46	4857	.75	2.06	.50	4954	.71
Canada (Eng.)	2.56	.30	9313	.63	2.31	.44	9261	.75	2.13	.52	9221	.77
Canada (Fr.)	2.38	.41	3400	.64	n.c.	n.c.	n.c.	n.c.	2.06	.55	3392	.75
England	2.50	.37	3614	.67	2.13	.51	3568	.74	2.13	.47	3620	.71
Finland	2.44	.35	3593	.56	1.94	.51	3542	.77	1.75	.50	3590	.73
Ghana	2.69	.27	489	.48	2.75	.27	488	.65	2.38	.40	488	.50
Hong Kong (Form 6)	n.c.	n.c.	n.c.	n.c.	2.38	.45	5903	.69	2.38	.42	5918	.60
Hong Kong (Form 7)	n.c.	n.c.	n.c.	n.c.	2.38	.45	3483	.70	2.31	.43	3525	.60
Hungary	2.56	.25	2018	.44	2.00	.49	2005	.72	1.81	.42	2018	.62
Italy	2.56	.33	6682	.65	n.c.	n.c.	n.c.	n.c.	2.00	.46	6684	.71
Japan	2.25	.44	6556	.67	1.75	.55	6549	.82	2.00	.52	6553	.70
Korea	2.75	.21	8331	.48	1.88	.48	8331	.75	2.19	.49	8330	.67
Papua New Guinea	2.25	.43	780	.56	2.38	.46	779	.75	n.c.	n.c.	n.c.	n.c.
Poland	2.50	.25	3240	.46	2.13	.43	3228	.71	1.88	.45	3228	.67
Singapore	2.50	.32	3552	.53	2.13	.45	3522	.72	2.25	.46	3548	.69
Sweden	2.50	.39	3273	.65	2.38	.43	2119	.70	2.06	.57	3273	.79
Thailand	2.63	.29	7120	.39	2.25	.45	7107	.72	2.55	.41	7120	.52

n.c. = not calculated because of missing data

Appendix G

Population 3 ctd.

Country	Non-harmful Aspects of Science				Career Interest in Science				Facility of Learning in Science			
	M	sd	N	α	M	sd	N	α	M	sd	N	α
Australia	2.13	.53	4956	.70	1.94	.49	4956	.64	2.00	.40	4845	.74
Canada (Eng.)	2.25	.47	9292	.70	2.25	.44	9307	.65	2.13	.43	9198	.77
Canada (Fr.)	1.94	.50	3398	.65	2.25	.48	3396	.65	n.c.	n.c.	n.c.	n.c.
England	2.31	.52	3618	.72	1.94	.48	3624	.65	2.00	.44	3570	.76
Finland	2.38	.45	3578	.62	1.88	.45	3589	.61	2.06	.43	3512	.76
Ghana	1.88	.43	488	.44	2.44	.35	487	.47	2.25	.49	479	.84
Hong Kong (Form 6)	2.00	.49	5915	.59	2.25	.39	5916	.44	2.00	.43	5872	.76
Hong Kong (Form 7)	2.00	.50	3523	.60	2.25	.40	3523	.45	2.06	.41	3460	.72
Hungary	2.25	.44	2018	.60	2.13	.38	2018	.52	n.c.	n.c.	n.c.	n.c.
Italy	2.19	.52	6685	.77	2.38	.40	6677	.61	2.06	.45	6531	.79
Japan	2.00	.50	6557	.66	1.94	.50	6550	.69	1.69	.49	6545	.82
Korea	1.88	.48	8329	.56	2.38	.38	8328	.47	1.88	.37	8325	.66
Papua New Guinea	1.81	.50	779	.55	2.06	.43	778	.46	1.81	.43	776	.70
Poland	2.06	.47	3240	.65	2.38	.32	3241	.49	2.00	.41	3215	.73
Singapore	1.94	.48	3552	.62	2.06	.41	3546	.54	2.00	.41	3474	.73
Sweden	2.00	.56	3274	.70	2.19	.48	3278	.67	2.13	.48	2134	.78
Thailand	2.00	.49	7120	.51	2.56	.34	7119	.44	1.94	.50	7105	.80

n.c. = not calculated because of missing data

Population 3 ctd.

Country	Teacher Directed Learning				Student Participation				Practical Work			
	M	sd	N	α	M	sd	N	α	M	sd	N	α
Australia	2.06	.39	4798	.68	1.50	.34	4779	.61	2.50	.40	4781	.73
Canada (Eng.)	2.25	.36	9201	.66	1.44	.31	9159	.57	2.56	.34	9179	.61
Canada (Fr.)	2.13	.40	3259	.69	1.50	.35	3252	.55	2.38	.40	3261	.68
England	2.00	.36	2879	.63	1.38	.30	2872	.59	2.38	.32	2871	.51
Finland	1.88	.34	3244	.52	1.25	.29	3171	.59	2.56	.44	3179	.74
Ghana	2.31	.38	489	.71	1.75	.38	471	.60	1.38	.30	488	.51
Hong Kong (Form 6)	2.19	.48	5915	.61	1.69	.45	5910	.51	2.56	.39	5914	.76
Hong Kong (Form 7)	2.19	.48	3522	.64	1.75	.45	3521	.54	2.75	.25	3523	.59
Hungary	2.25	.35	2019	.63	1.44	.32	2019	.65	2.13	.41	2019	.74
Italy	2.19	.38	6542	.33	1.69	.54	6361	.66	1.69	.56	6396	.73
Japan	1.81	.32	6553	.66	1.44	.30	6542	.54	2.38	.39	6548	.74
Korea	2.06	.40	8312	.66	1.50	.38	8304	.66	2.00	.55	8309	.83
Papua New Guinea	2.38	.31	778	.54	1.56	.31	775	.52	2.50	.30	776	.55
Poland	2.06	.37	3238	.66	1.50	.37	3238	.73	2.00	.39	3239	.63
Singapore	2.00	.36	3517	.65	1.44	.32	3497	.53	2.63	.33	3501	.46
Sweden	2.00	.30	2087	.54	1.44	.27	2087	.56	2.38	.30	2090	.38
Thailand	2.31	.32	7117	.55	1.81	.32	7107	.49	2.63	.28	7119	.46

Country	Homework Effort				Classroom Effort			
	M	sd	N	α	M	sd	N	α
Finland	1.81	.72	3221	.85	2.13	.50	3215	.66
Hong Kong (Form 6)	-	-	-	-	2.31	.43	5887	.56
Hong Kong (Form 7)	-	-	-	-	2.25	.45	3505	.57
Hungary	2.25	.34	2017	.65	2.19	.36	2017	.64
Italy	2.25	.44	6663	.75	2.25	.39	6255	.63
Japan	-	-	-	-	2.06	.52	6624	.67
Korea	1.63	.88	7986	.91	2.38	.42	7987	.54
Poland	2.25	.39	3239	.69	2.13	.38	3239	.62
Singapore	2.13	.43	3386	.70	2.13	.46	3376	.61
Thailand	2.31	.33	6133	.49	2.44	.41	6963	.42

Appendix H

Offering and requirement patterns [single frequencies and missing data omitted] with percent schools covered (by country) (Population 2)

Australia	Percent Schools:	94.3
Offerings	Requirements	Frequency
General	General	164
General	None	31
GBCP	GBCP	2

Canada (Eng.)	Percent Schools:	97.5
Offerings	Requirements	Frequency
General	General	105
General	None	38
Biology	Biology	6
Physics	Physics	5
Biology	None	2
G B	None	2

Canada (Fr.)	Percent Schools:	93.4
Offerings	Requirements	Frequency
Biology	Biology	25
G B	Biology	11
General	General	10
Biology	None	7
B P	B	7
General	None	5
B P	None	5
G B	None	5
GB	GB	4
GBP	GBP	2
E B	E B	2
E B	None	2

England	Percent Schools:	92.6
Offerings	Requirements	Frequency
BCP	BCP	68
G BCP	G BCP	21
General	General	18
G BCP	None	7
BCP	None	4
B CP	B	4
BP	BP	2
GB	GB	2

Finland	Percent Schools:	100.0
Offerings	Requirements	Frequency
EBCP	EBCP	89

Hong Kong	Percent Schools:	99.1
Offerings	Requirements	Frequency
General	General	100
General	None	4
BCP	BCP	2

Hungary	Percent Schools:	100.0
Offerings	Requirements	Frequency
EBCP	EBCP	97

Italy (Grade 8)	Percent Schools:	93.4
Offerings	Requirements	Frequency
General	General	126
GEBCP	GEBCP	13
ECP	ECP	8
General	None	4
EBCP	EBCP	4
GE	GE	3
GEBP	GEBP	3
GCP	GCP	2
EBP	EBP	2
GECP	GECP	2
BCP	BCP	2

Italy (Grade 9)	Percent Schools:	78.0
Offerings	Requirements	Frequency
General	General	9
GP	GP	5
Physics	Physics	3
EP	EP	3
Chemistry	Chemistry	2
Earth Science	Earth Science	2
GBCP	GBCP	2
GEP	GEP	2
EBP	EBP	2
GECP	GECP	2

Japan	Percent Schools:	100.0
Offerings	Requirements	Frequency
General	General	199

Korea	Percent Schools:	99.5
Offerings	Requirements	Frequency
General	General	187

Netherlands	Percent Schools:	84.0
Offerings	Requirements	Frequency
ECP	ECP	21
Physics	Physics	18
EBCP	EBCP	15
EC B P	EC	14
EBC P	EBC	11
E B	None	11
BCP	BCP	10
EB	EB	6
EP	EP	5
E B C P	E	5
Biology	None	4
Biology	Biology	4
ECP	None	3
EB	None	3
E B	Biology	3
E B P	E P	3
EBCP	None	3
Physics	None	2
E P	Physics	2
B P	BP	2
EBP	EBP	2
EB P	None	2
BCP	None	2
EBC	EBC	2
C P	Physics	2
E B P	None	2

Offering and Requirement Patterns

Nigeria	Percent Schools:	88.7
Offerings	Requirements	Frequency
BCP	BCP	24
BCP	Biology	18
BCP	None	15
BCP	None	4
B CP	None	2

Papua New Guinea	Percent Schools:	100.0
Offerings	Requirements	Frequency
General	General	74

Philippines	Percent Schools:	98.0
Offerings	Requirements	Frequency
Chemistry	Chemistry	209
Chemistry	None	21
Biology	Biology	10
Biology	None	2
EC	EC	2

Poland	Percent Schools:	100.0
Offerings	Requirements	Frequency
GEBCP	GEBCP	198

Singapore	Percent Schools:	98.2
Offerings	Requirements	Frequency
Biology	None	62
B CP	None	40
CP	None	25
CP	None	13
BCP	None	12
BP	None	3
BCP	None	3
BC P	None	3
BCP	BC	2

Thailand	Percent Schools:	100.0
Offerings	Requirements	Frequency
General	General	92

United States (Phase 1)	Percent Schools:	89.0
Offerings	Requirements	Frequency
General	General	16
General	None	11
G B	None	10
G E B	None	7
Earth Science	Earth Science	6
Biology	None	4
Earth Science	None	3
G B	G B	2
E B	None	2
G E	None	2
G E B C	None	2

The notation used in the pattern description designates the letters G, E, B, C, and P to denote General Science, Earth Science, Biology, Chemistry, and Physics respectively. When the letters do not have spaces between them, the grouped subjects have identical pupils enrolled. If the letters are spaced, different groups of pupils are enrolled in each. When a group of subjects is required all pupils in the grade experience all of the designated subjects. When not required, this implies a 'track' or 'stream' within which specific subgroup of all pupils in the grade experience all of the designated subjects. In Finland E represents Geography.

Appendix I

The Partial Least Squares Analysis

The analyses reported in Chapter 8 are based on Partial Least Squares estimation method. PLS uses least squares estimates and because distributional assumptions are not required, it is very robust. In terms of educational studies the best description of the PLS path technique has been written by Norbert Sellin in Appendices B and C to 'The IEA Classroom Environment Study'[1] and readers are referred to that publication.

The analyses have been undertaken at the between student overall level. Strictly speaking, it would have been desirable to undertake the analyses at two levels: the within class/school and between class/school levels. On the other hand, there were temporal constraints and it was felt that selecting all path coefficients above the .10 level would allow those school and teacher variables with an influence to be identified.

First, two tables are presented in this appendix: Table I.1 presents the loading for two or more manifest variables which have been included in a construct. The loading for one variable indicated the relative strength of that variable versus other variables in the construct. Table I.2 presents the direct and total effects between construct effects (the path coefficients) as well as the R^2 of all constructs predicting a particular construct. The path model used has been described in Chapter 8.

As an example of a construct take the socio-economic status of the home for Korea. There are four manifest variables: family size, father's education, mother's education, and father's occupation. Their respective loadings are -0.63, 0.83, 0.32, and 0.73. It will be seen that the family size is related -0.63 to the construct as a whole indicating that children from larger families have mothers and fathers with fewer years of education. It is, however, the father's education which dominates the construct.

Table I.2 presents the estimated direct and total effects between constructs as well as the R^2 prediction.

The indirect effect is the total minus the direct effect. Take the 'Like School' construct and, take Singapore. It will be seen that the total effect of the 'Socio-economic Status of the Home' is 0.10 made up of a direct effect of 0.05 and an indirect effect of 0.05 (working through the 'Literacy of the Home'). The 'Literacy of the Home' has a path of 0.13, 'Sex of Student' 0.10 (i.e. girls like school more than boys) and 'School Equipment' 0.01. The four constructs have an R^2 of .035 indicating that only 3.5 percent of the variance of 'Like School' is explained by them.

It is for the reader to examine the predictions of constructs in which he or she is interested.

It is likely that readers would like to postulate other models and test them using PLS or some other statistical technique.

---

1 Anderson, L. W., Ryan, D. W. and Shapiro, B. J. (Eds.) 1989 *The IEA Classroom Environment Study*. Pergamon, Oxford.

TABLE I.1: Summary of constructs and associated manifest variables (Population 2)

	Canada (Fr.)	China	Hong Kong	Korea	Philippines	Poland	Singapore	Pooled PLS loadings	Component weights
Socio-economic Status of Home									
Family Size	1.00	-.59	-.55	-.63	-.20	-.74	-.53	-.39	-.43
Father's Education	-.07	.87	.69	.83	.74	.89	-	.65	.79
Mother's Education	-	.97	.90	.32	.77	.79	.89	.60	.71
Father's Occupation	-	.91	-	.73	.90	-	.55	.90	.67
Literacy of Home									
Word Knowledge	-	-	-	1.00	.32	-	.78	.48	.42
Books in Home	-	1.00	.80	-	.83	.95	.71	.81	.76
Use of Dictionary	-	-	.75	-	.71	.56	.42	.62	.72
Teacher Training and Experience									
Years Post-Secondary Education	.54	1.00	.88	.53	.49	.37	.28	.66	.72
Years Post-Sec. Science Educ.	.86	-	.76	.69	.68	.70	.82	.75	.73
Years Teaching Experience	-.45	-	.38	-.25	.64	.18	.38	.44	.12
Days In-Service Training	.06	-	.31	.77	.49	.81	.03	.36	.53
School Equipment									
Number of Labs	.99	.42	.83	.79	.68	.75	.63	.74	.75
Use of Labs (Percent Time)	.22	-	.84	.07	.67	.92	.27	.61	.52
Lab Assistants	.54	.99	.84	.79	.41	-	.44	.68	.74
Like School									
School Is Challenging	.50	-.05	.47	.42	.56	.19	.50	.48	.32
School Is Not Enjoyable (r)	.71	-.27	.62	.52	.50	.43	.51	.52	.56
Enjoy Everything at School	.54	.67	.44	.25	.37	.56	.23	.36	.51
Bored at School (r)	.68	.56	.59	.49	.54	.66	.67	.60	.61
Do Not Like Many Subjects (r)	.72	.59	.57	.60	.55	.60	.58	.59	.60
School Most Enjoyable	.23	.56	.44	.41	.39	.60	.40	.41	.51
Want More Education	.61	.40	.57	.59	.49	.28	.46	.50	.39
Dislike Schoolwork (r)	.71	.63	.55	.68	.61	.74	.61	.64	.65
Teaching Style									
Teacher Reviews First	.47	.48	.48	.27	.57	.53	.50	.39	.40
Teacher Starts With Explanation	.47	.52	.51	.76	.50	.61	.57	.43	.41
Teacher Finished With Summary	.49	.63	.58	.19	.40	.45	.57	.51	.51
Students Choose Topic of Study	.23	.52	.26	.15	-.67	.26	.21	.31	.29
Teacher Uses Student Ideas	.47	.59	.40	.33	-.37	.34	.31	.42	.42
Students Set Own Problems; Teacher Helps	.61	.52	.55	.38	-.11	.67	.57	.58	.55
Teacher Sets Problems; Students Solve	.44	-	.50	.36	-.04	.58	.23	.48	.48
Students Set Own Problems and Solve	.49	.51	.38	.30	-.19	.47	.30	.48	.52

TABLE I.1 (ctd.): Summary of constructs and associated manifest variables (Population 2)

	Canada (Fr.)	China	Hong Kong	Korea	Philippines	Poland	Singapore	Pooled PLS loadings	Component weights
Experiments									
Students Do Experiments	.86	.80	.87	.62	.72	.31	.77	.80	.81
Small Group Experiments	.88	.77	.80	.86	.88	.99	.80	.82	.81
Classroom Effort									
Teacher Asks You a Question	-	.64	-	-.13	.51	.28	.32	.26	.38
Teacher Asks Someone Else a Question	-	.50	-	.14	.43	.50	.53	.42	.51
Pay Attention	-	.53	-	.65	.53	.66	.67	.60	.57
Make Up for Missed Lessons	-	.61	-	.44	.50	.67	.55	.54	.58
Deliberately Miss a Class	-	-	-	.77	.66	.58	.61	.68	.54
Homework									
Hours Homework per Week in All Subjects	.97	.95	.97	.99	.99	.99	.90	.97	.85
Hours Homework per Week in Science	.74	.84	.66	.18	.56	.53	.81	.67	.85
Time									
Hours Science Instruction per Week	.40	-	.30	.98	.38	.93	.38	.60	.55
Amount of Practical Work	-.09	1.00	.44	-.25	.05	.56	.32	.34	.67
Total Hrs. Teaching Instruction	.91	-	.74	-	.89	-	.89	.80	.62
Views of and Interest in Science									
Science Important for Country Development	.41	.45	.48	.33	.45	.22	.37	.42	.44
Worth Spending Money on Science	.52	.49	.41	.55	.37	.38	.50	.45	.41
Inventions Improve Living Standard	.49	.42	.55	.40	.57	.37	.45	.51	.49
Science Is for Creative People	.19	.49	.48	.46	.53	.39	.39	.46	.44
Science Helps You Learn More About World	.44	.52	.42	.35	.61	.30	.33	.47	.45
All Students Should Learn Science	.09	.13	.37	.28	.50	.02	.38	.34	.36
Science Is an Enjoyable Subject	.72	.59	.60	.60	.56	.65	.67	.61	.64
School Science Is Interesting	.67	.56	.58	.41	.56	.64	.54	.55	.58
Science Is a Difficult Subject	.52	-.06	.19	.42	.13	.43	.40	.28	.30
Science Is Difficult/Calculations	.37	.18	.21	.33	-.06	.39	.27	.18	.17
Science Is Difficult/Apparatus	.39	.15	.37	.28	.17	.31	.40	.31	.27
Science Is Relevant for Everyday	.39	.59	.57	.33	.53	.45	.49	.50	.51

Appendix I

TABLE I.2: Direct and total between construct effects and R-square (Population 2)

Predicted/Predictor	Canada (Fr.) D	T	China D	T	Hong Kong D	T	Korea D	T	Philippines D	T	Poland D	T	Singapore D	T	Pooled Direct Effects PLS	Regression
Literacy of Home																
Socio-economic Status of Home	-	-	.41	.41	.41	.41	.25	.25	.51	.51	.43	.43	.40	.40	.43	.44
R^2	-		.168		.168		.063		.260		.185		.160			
Like School																
Socio-economic Status of Home	-.08	-.28	-.08	-.06	.06	.08	.02	.06	.10	.17	-.08	-.06	.05	.10	.04	-.01
Literacy of Home	.19	.19	.05	.05	.12	.12	.14	.14	.14	.14	.04	.04	.13	.13	.12	.07
Sex of Student	.02	.02	.07	.07	.08	.08	.19	.19	.16	.16	.15	.15	.10	.10	.13	.12
School Equipment	.02	.02	-.14	-.14	.07	.07	-.02	-.02	.16	.16	-.05	-.05	.01	.01	.03	-.01
R^2	.041		.036		.036		.055		.102		.031		.035			
School Science Equipment																
Location of School	-.20	-.20	-	-	-	-	-.06	-.06	-.01	-.01	-.10	-.10	.04	.04	-.06	-.09
Size of School	.67	.67	.73	.73	.48	.48	.45	.45	.43	.43	.43	.43	.42	.42	.40	.35
R^2	.569		.527		.227		.246		.193		.246		.171			
Experiments																
Teacher Training and Experience	.12	.12	.10	.10	.19	.19	-		.07	.07	.06	.06	.10	.10	.07	.08
R^2	.014		.010		.036				.005		.004		.010			
Homework																
Socio-economic Status of Home	0	0	.04	.04	.06	.06	.08	.11	.03	.08	-.05	.02	.11	.20	.03	.02
Literacy of Home	0	-	.15	.15	.13	.15	.18	.18	.11	.11	.15	.15	.21	.21	.15	.11
Sex of Student	-.02	-.02	.02	.02	.07	.07	.02	.02	.03	.03	0	0	.03	.03	.04	.01
Teacher Training and Experience	.08	.14	.02	.02	.07	.07	-		.16	.16	-.05	-.05	.10	.10	.04	.01
R^2	.024		.030		.062		.030		.018		.031		.100			
Classroom Effort																
Sex of Student	-	-	.01	.03	-	-	-.03	.02	.07	.12	.14	.18	.05	.07	.06	.06
Like School	-	-	.22	.22	-	-	.26	.26	.31	.31	.27	.27	.25	.25	.25	.24
R^2	-		.048		-		.064		.108		.101		.690			

D direct effect
T total effect

TABLE I.2 (ctd.): Direct and total between construct effects and R-square (Population 2)

Predicted/Predictor	Canada (Fr.) D	T	China D	T	Hong Kong D	T	Korea D	T	Philippines D	T	Poland D	T	Singapore D	T	Pooled Direct Effects	PLS Regression
Time																
Teacher Training and Experience	.11	.11	.09	.09	.17	.17	-.26	-.26	.09	.09	.33	.33	.27	.27	.14	.16
School Equipment	.30	.30	.25	.25	.69	.69	.11	.11	.50	.50	.31	.31	.27	.27	.36	.40
R^2	.109		.077		.569		.078		.272		.238		.131			
Views of and Interest in Science																
Socio-economic Status of Home	-.03	-.06	.09	.07	.06	.08	.11	.13	.07	.15	.08	.08	.11	.15	.08	.10
Sex of Student	-.14	-.06	.01	.03	-.16	-.13	-.17	-.10	-.01	-.08	-.02	-.	-.14	-.09	-.08	-.07
Teacher Training and Experience	.05	.07	-.01	-.00	.06	.10	0	0	.03	.03	0	0	.05	.06	.03	.01
Experiments	.16	.16	.06	.06	.21	.21	-.01	-.01	.06	.06	.11	.11	.11	.11	.09	.08
Teaching Style	.05	.05	.13	.13	.07	.07	.13	.13	.09	.09	0	0	.06	.06	.07	.08
Like School	.39	.39	.37	.39	.34	.34	.34	.37	.45	.45	.34	.37	.33	.38	.36	.39
Classroom Effort	-	.10	-	.10	-	.	-	.12	-	.14	-	.12	-	-.17	.14	.13
R^2	.213		.212		.231		.205		.335		.185		.241			
Achievement																
Socio-economic Status of Home	-.08	-.10	.12	.13	.19	.15	.06	.04	.12	.25	.07	.17	.21	.23	.11	.13
Literacy of Home	-	.13	.15	.19	.04	.07	.41	.43	.17	.19	.19	.17	.18	.23	.18	.24
Sex of Student	-.26	-.27	.01	-.26	-.16	-.19	-.19	-.20	-.12	-.09	-.02	-.16	-.18	-.19	-.17	-.14
School Equipment	-.01	.02	-.26	.01	.07	.17	0	0	.10	.13	0	0	.01	.04	.03	-.02
Teacher Training and Experience	.11	.15	-.07	-.06	.08	.17	.05	-.07	.13	.14	.02	0	.11	.18	.05	.07
Experiments	.12	.16	.07	.07	.16	.22	-.07	-.07	.08	.09	.02	.02	.10	.12	.09	.09
Teaching Style	-.18	-.16	-.06	-.05	-.14	-.12	.02	.05	.14	.15	-.12	-.10	-.14	-.13	-.11	-.13
Homework	.08	.08	.12	.12	.14	.14	.04	.04	.06	.06	.11	.11	.14	.14	.09	.07
Like School	-.02	.08	-.03	.07	-.03	.06	-	-	.12	.12	.06	.06	.08	.09	-	-
Classroom Effort	-	.10	-	.11	.17	.17	.07	.09	.11	.13	.08	.08	.08	.12	.11	.05
Time	-	.02	.02	.02	.17	.17	.02	.02	.03	.03	.03	.03	.11	.11	.08	.13
Views of and Interest in Science	.05	.05	.02	.02	.27	.27	.20	.20	.17	.17	.12	.12	.19	.19	.20	.19
R^2	.243		.199		.343		.401		.318		.134		.452			

D   direct effect
T   total effect

Secondly, Tables I.3 and I.4 present the data for Population 1. It will be noted that one or two of the constructs are slightly different from those for Population 2.

The analyses were run separately for biology, chemistry, physics. The Tables I.5 and I.6 present the manifest variables' loadings within each construct for each country for biology, and the direct and total between construct effects of biology respectively. The Tables I.7 and I.8 do the same for chemistry and Tables I.9 and I.8 for physics.

From Table 1.4 it can be seen that at Population 1 level, the home has a large effect on achievement. Like school has some effect and this, in turn, is mostly predicted by Views of and Interest in Science. Where teachers use a student-centered style of teaching this has a negative effect on unknowns.

At Population 3 level, the effect of home on achievement is negligible except for Korea. This indicates that the students still in school at this level are from a socially homogeneous set of homes. Korea has a higher percentage of an age group in school at this level (see Table 1.2 in Chapter 1). There are no consistent results for other variables. Variation in school equipment has an influence on achievement in Hong Kong, Korea and Singapore. This, in turn, is related to size and rurality of school. Again, a student-centered approach to teaching is negatively associated with achievement and this is not easy to explain. Variance in teacher training has an effect in Hong Kong. With the exception of Korea, time spent on homework has a positive effect on achievement. The number of hours of instruction in the subject area has a positive effect on achievement for biology and chemistry, but in physics there would appear to be little variance within a country on time allocated to the teaching of physics. Views of and Interest in Science always have a positive effect on achievement. These attitudes are influenced mostly by 'Like School' and 'Classroom Effort.'

It would seem that further detailed analyses within each country are required in each of the countries before definitive statements can be made.

The Partial Least Squares Analysis

TABLE I.3: Construct Loadings (Population 1)

	Canada (Fr.)	Hong Kong	Korea	Philippines	Poland	Singapore
Socio-economic Status of Home						
Family Size	1.00	1.00	-.65	-.13	-.17	-.65
Mother's Education	-	-	.81	-	.84	-
Father's Education	-	-	-	-	-	-
Father's Occupation	-	-	.58	.99	-	.85
Literacy of Home						
Word Knowledge Test	-	-	.81	-.41	-	.83
No. of Books in Home	-	.66	.68	.81	.97	.65
Use of Dictionary	-	.85	.48	.55	.47	.30
Teacher Training and Experience						
Years Post-Secondary Education	.70	.43	.98	.54	.69	-.11
Years Post-Sec Science Education	.58	.14	.12	.41	.90	.39
Years Teaching Experience	-.72	.15	.01	.23	.02	-.21
Days In-Service Training	.25	.93	.24	.81	.57	.97
Like School						
Dislike Schoolwork (r)	.68	.56	.54	.60	.66	.58
School is not Enjoyable (r)	.74	.61	.59	.63	.42	.62
Want More Education	.48	.60	.35	.58	.49	.54
Enjoy Everything at School	.57	.57	.16	.48	.39	.31
Bored at School (r)	.67	.61	.66	.50	.68	.67
Do Not Like Many Subjects (r)	.70	.61	.53	.60	.70	.61
School Most Enjoyable	.20	.47	.30	.47	.34	.28
Experiments						
Students Do Experiments	.77	.85	.69	.57	.65	.35
Small Group Experiments	.85	.71	.87	.92	.83	.89
Teaching Style						
Student Uses Textbook	.00	-.04	-.35	.28	.45	-.78
Student Uses Library Books	.44	.59	.48	-.07	-.37	.28
Student Selects Topics for Study	.68	.65	.58	-.35	-.77	.46
Teacher Uses Student Ideas	.61	.60	.57	.05	-.60	.30
Student Copies Notes from Blackboard	.53	.45	-.06	.21	.01	-.05
Student Has Regular Science Tests	.10	.31	.47	.89	-.12	-.45
School Science Equipment						
Number of Labs	.15	.42	.81	.19	.50	.74
Use of Labs	.80	.93	-.35	.96	.97	.99
Time						
Hours Science Instruction per Week	.58	.77	.99	.79	.99	.75
Percentage Practical	.81	.37	.10	.40	.25	.64
No. of Allocated Hours of Science	-.11	.60	-	-.51	-	.35
View of and Interest in Science						
Science Is An Enjoyable Subject	.65	.55	.50	.48	.53	.46
Instructions Improve Standards of Living	.34	.46	.43	.58	.49	.59
Science Ruins The Environment	.35	.10	.28	.52	.55	.43
School Science Is Interesting	.70	.64	.52	.61	.48	.45
Science Helps Improve World's Future	.47	.51	.44	.62	.42	.51
Discoveries Do Harm (r)	.45	.25	.33	.45	.60	.46
Science Important for Country Development	.42	.52	.42	.55	.45	.52
Spend More on Science Research	.38	.17	.46	.38	.23	.42
Science Lessons Are Interesting	.59	.57	.48	.39	.44	.39

Appendix I

TABLE I.4: Direct and total between construct effects and R-square (Population 1)

Predicted/Predictor	Canada (Fr) D	Canada (Fr) T	Hong Kong D	Hong Kong T	Korea D	Korea T	Philippines D	Philippines T	Poland D	Poland T	Singapore D	Singapore T
Literacy of Home												
Socio-economic Status	-	-	.02	.02	.37	.37	.30	.30	.37	.37	.24	.24
R^2	-		.000		.138		.089		.137		.055	
Like School												
Socio-economic Status	-.05	-.05	-.09	-.09	.02	.09	.14	.17	.03	.05	.02	.05
Literacy of Home	-	-	.08	.08	.18	.18	.09	.10	.06	.06	.14	.14
Sex	.20	.20	.12	.12	.11	.11	.08	.08	.14	.14	.06	.06
R^2	.043		.030		.048		.043		.024		.024	
School Science Equipment												
Location of School	.05	.05	-	-	-.13	-.13	.01	.01	-.21	-.21	-.15	-.15
Size of School	-.15	-.15	.39	.39	.26	.26	.19	.19	.41	.41	.05	.05
R^2	.029		.151		.128		.036		.318		.026	
Experiments												
Teacher Training	.15	.15	-.03	.03	-.02	-.02	.05	.05	-.02	-.02	-.04	-.04
R^2	.023		.001		000		.002		.000		.000	
Time												
Teacher Training	.25	.25	.29	.29	-.09	-.09	.16	.16	.58	.58	.22	.22
Location	-	.01	-	-	-	.01	-	.00	-	.04	-	-
Size of School	-	-.03	-	.04	-	-.01	-	.02	-	.08	-	-
Science Equipment	.19	.19	.09	.09	-.04	-.04	.10	.10	.20	.20	.02	.02
R^2	.116		.090		.011		.041		.427		.051	
Views of and Interest in Science												
Socio-economic Status	-.02	-.02	-.06	-.09	.06	.08	.05	.15	.02	.04	.08	.10
Literacy of Home	-	-	-	.03	-	.05	-	.06	-	.02	-	.05
Sex	-.09	-.02	-.08	-.03	-.10	-.07	.02	.06	-.01	.06	-.09	-.07
Teacher Training	-.04	-.03	.03	.03	.02	.02	.02	.02	-.01	-.01	.04	.04
Like School	.35	.35	.34	.34	.27	.27	.58	.58	.44	.44	.34	.34
Experiments	.07	.07	.06	.06	.16	.16	.04	.04	.03	.03	.04	.04
Teacher Style	.03	.03	.11	.11	.04	.04	-.11	-.11	-.07	-.07	-.08	-.08
R^2	.128		.155		.123		.402		.207		.148	
Achievement												
Socio-economic Status	-.08	-.09	-.14	-.17	.02	.21	.10	.19	.10	.16	.18	.28
Literacy of Home	-	-	.18	.18	.48	.50	.10	.13	.16	.16	.33	.35
Sex	0	0	-.13	-.13	-.17	-.17	-.05	-.03	-.15	-.13	-.13	-.13
Teacher Training	-.03	-.04	-	.01	-	0	-	.01	-	0	-	.01
Like School	.06	.10	.08	.15	.06	.09	.20	.31	.08	.13	.14	.19
Experiments	-.03	-.02	-.09	-.08	.05	.07	.04	.05	-.12	-.11	.01	.02
Teaching Style	-.15	-.15	-.16	-.14	-.13	-.12	-.15	-.17	-.22	-.23	-.17	-.18
Views of and Interest in Science	.12	.12	.20	.20	.11	.11	.19	.19	.10	.10	.16	.16
R^2	.057		.155		.363		.232		.182		.355	

D   direct effect
T   total effect

TABLE I.5: Summary of constructs and associated manifest variables (Population 3 - Biology)

	Canada (Fr)	Hong Kong (Form 6)	Hong Kong (Form 7)	Korea	Poland	Singapore
Socio-economic Status of Home						
Family Size	1.0	.81	.91	-.42	-.25	-
Father's Education	-	.34	.14	.95	.88	-
Mother's Education	-	.29	.16	.87	.93	-
Father's Occupation	-	-	-	.52	-	1.0
Literacy of Home						
Word Knowledge Test	-	-	-	.79	-	.58
No. of Books in Home	-	.82	.88	.73	.94	.83
Use of Dictionary	-	.71	.63	.28	.61	.46
Teacher Training and Experience						
Years Post-Secondary Education	-.23	.79	.64	.61	.17	-
Years Post-Sec Science Education	-.47	.18	.14	.51	.88	-
Years Teaching Experience	.75	.07	.70	.49	.22	-
Days In-Service Training	.74	.88	.63	.54	.49	-
School Equipment						
Number of Labs	.66	-.43	.98	.99	.99	.69
Use of Labs (Percent Time)	.70	.99	.15	.29	.13	.49
Lab Assistants	-	-	-	-	-	-
Like School						
School is not Enjoyable (r)	.33	.67	.63	.57	.43	.68
Enjoy Everything at School	.72	.34	.49	.48	.43	.38
Bored at School (r)	.62	.71	.72	.64	.72	.64
Do Not Like Many Subjects (r)	.69	.62	.44	.53	.42	.64
School Most Enjoyable	.50	.41	.35	.60	.46	.55
Want More Education	.59	.27	.25	.45	.27	.32
Dislike Schoolwork (r)	.51	.62	.63	.70	.80	.74
Experiments						
Students Do Experiments	.70	.93	.80	.75	.76	.64
Small Group Experiments	.98	.79	.69	.80	.83	.80
Teaching Style						
Teacher Reviews First	.33	.54	.56	.50	.53	.31
Teacher Starts with Explanation	.36	.56	.65	.55	.56	.67
Teacher Finished with Summary	.66	.59	.59	.59	.66	.56
Teacher Uses Students' Ideas	.30	.51	.52	.58	.20	.46
Students Set Own Problems; Teacher Helps	.29	.51	.58	.43	.72	.39
Teacher Sets Problems; Students Solve	.62	.50	.46	.52	.47	.34
Students Set Own Problems and Solve	.56	.50	.42	.37	.38	.43
Classroom Effort						
Teacher Asks You a Question	-	.47	.37	.43	.16	.33
Teacher Asks Someone Else a Question	-	.58	.42	.53	.48	.52
Pay Attention	-	.61	.62	.64	.74	.59
Make Up for Missed Lessons	-	.60	.69	.63	.50	.71
Deliberately Miss a Class	-	.59	.60	.60	.62	.66
Homework						
Hours Homework per Week in all Subjects	.92	.95	.97	.99	.87	.94
Hours Homework in Science	.95	.94	.97	.58	.88	.95
Time						
Hours Science instruction per week	.85	-.05	-.50	-.12	.44	-
Amount of Labwork by Teacher	.60	.83	.58	-.44	.72	-
Amount of Practical Work	.09	.40	.15	.46	.68	-
Total Hours Teaching Instruction	.31	.82	.81	.70	-	1.0
View of and Interest in Science						
Worth Spending Money on Science	.41	.43	.35	.24	.43	.34
Anxiety in Modern Society Due Science	.23	.21	.14	.19	.23	.14
People Understand Science Better Off	.57	.10	.17	.26	.16	.26
Inventions Improve Living Standards	.60	.09	.19	.12	.20	.18
Science is for Creative People	.57	.27	.22	.16	.33	.26
Science Causes World's Problems	.53	.24	.11	.21	.18	.29
Biology is an Enjoyable Subject	-	.56	.51	.70	.60	.52
School Biology Interesting	-	.22	.66	.74	.67	.69
Science Lessons Interesting	-	.65	.64	.46	.60	.47
Like Science	-	.54	.50	.53	.54	.52

Appendix I

TABLE I.6: Direct and total between construct effects and R-square (Population 3 - Biology)

Predicted/Predictor	Canada(Fr) D	T	Hong Kong (Form 6) D	T	Hong Kong (Form 7) D	T	Korea D	T	Poland D	T	Singapore D	T
Literacy of Home												
Socio-economic Status of Home	-	-	.08	.08	.05	.05	.45	.45	.46	.46	.15	.15
R^2	-		.007		.002		.203		.213		.022	
Like School												
Socio-economic Status of Home	.04	.04	-.09	-.09	-.02	-.02	-.10	-.04	-.10	-.07	-.06	-.06
Literacy of Home	-	-	.11	.11	.06	.06	.14	.14	.06	.06	.02	.02
Sex of Student	.16	.16	.12	.12	.04	.04	.16	.16	.06	.06	-.17	-.17
School Equipment	.02	.02	.01	.01	.03	.03	-.02	-.02	-.04	-.04	.11	.11
R^2	.032		.026		.005		.043		.04		.045	
School Science Equipment												
Location of School	-.13	-.13	-	-	-	-	-.33	-.33	-.08	-.08	-.39	-.39
Size of School	.40	.40	.13	.13	.24	.24	.34	.34	.35	.35	.37	.37
R^2	.201		.016		.059		.367		.148		.366	
Experiments												
Teacher Training and Experience	-.11	-.11	.14	.14	.04	.04	.09	.09	.01	.01	-	-
R^2	.012		.020		.001		.008		.000		-	
Homework												
Socio-economic Status of Home	.12	.12	.06	.07	.03	.03	.02	.04	.00	.06	.03	.05
Literacy of Home	-	-	.13	.13	.05	.05	.05	.05	.14	.14	.12	.12
Sex of Student	.36	.36	.54	.54	.79	.79	.16	.16	.17	.17	.11	.11
Teacher Training and Experience	-.15	-.15	.06	.06	.00	.00	.03	.03	.03	.03	-	-
R^2	.186		.358		.669		.030		.048		.031	
Classroom Effort												
Sex of Student	-	-	.02	.06	.04	.05	.08	.13	.11	.12	.11	.17
Like School	-	-	.31	.31	.34	.34	.34	.34	.24	.24	.37	.37
R^2	-		.098		.118		.128		.073		.162	
Time												
Teacher Training and Experience	.39	.39	.45	.45	.46	.46	.13	.13	.31	.31	-	-
School Equipment	.20	.20	.24	.24	.13	.13	.16	.16	-.08	-.08	-.42	-.42
R^2	.198		.300		.267		.046		.098		.176	
Views of and Interest in Science												
Socio-economic Status of Home	-.01	.00	.07	.03	.12	.11	-.02	-.03	.07	.06	.00	-.02
Sex of Student	-.16	-.14	-.08	-.04	-.19	-.17	-.10	-.04	-.04	.01	-.10	-.03
Teacher Training and Experience	-.03	-.03	.04	.05	-.03	-.03	.00	.00	-.03	-.03	-	-
Experiments	.01	.01	.07	.07	.04	.04	.03	.03	.07	.07	.13	.13
Teaching Style	.22	.22	.13	.13	.18	.18	.17	.17	.20	.20	.17	.17
Like School	.13	.13	.33	.37	.36	.39	.31	.34	.23	.23	.38	.40
Classroom Effort	-	-	.13	.13	.10	.10	.09	.09			.05	.05
R^2	.089		.208		.247		.166		.233		.231	
Achievement												
Socio-economic Status of Home	-.18	-.15	-.04	-.03	.03	.04	.01	.17	.05	.07	.03	.05
Literacy of Home	-	-	.01	.02	.03	.04	.37	.38	.01	.03	.16	.16
Sex of Student	-.02	.07	-.28	-.25	-.25	-.20	-.10	-.08	-.08	-.06	-.29	-.29
School Equipment	.09	.07	-.11	-.07	.09	.12	.13	.14	-.02	-.03	.29	.27
Teacher Training and Experience	-.03	-.14	.09	.18	.03	.14	.01	.04	.04	.04	-	-
Experiments	.15	.15	.07	.08	.06	.06	.00	.01	-.01	-.01	.08	.08
Teaching Style	-.20	-.16	-.12	-.11	-.09	-.08	-.16	-.13	-.12	-.11	-.03	-.02
Homework	.24	.24	.05	.05	.08	.08	.01	.01	.13	.13	.01	.01
Like School	.16	.18	-	.06	-	.03	-	.10	-	.03	-	.02
Classroom Effort	-	-	.08	.09	.06	.06	.14	.15	.01	.02	-.01	-.01
Time	-.12	-.12	.18	.18	.24	.24	.05	.05	.00	.00	.06	.06
Views of and Interest in Science	.18	.18	.11	.11	.03	.03	.17	.19	.08	.08	.06	.06
R^2	.217		.167		.149		.316		.049		.202	

D   direct effect
T   total effect

The Partial Least Squares Analysis

TABLE I.7: Summary of constructs and associated manifest variables (Population 3 - Chemistry)

	Canada (Fr)	Hong Kong (Form 6)	Hong Kong (Form 7)	Korea	Poland	Singapore
Socio-economic Status of Home						
Family Size	1.0	.82	.94	-.38	-.29	.99
Father's Education	-	.31	.09	.96	.80	-
Mother's Education	-	.27	.09	.87	.96	-
Father's Occupation	-	-	-	.58	-	-.25
Literacy of Home						
Word Knowledge Test	-	-	-	.79	-	.97
No. of Books in Home	-	.78	.84	.76	.92	.38
Use of Dictionary	-	.77	.70	.20	.41	.02
Teacher Training and Experience						
Years Post-Secondary Education	.89	.71	.68	-.11	-.05	-
Years Post-Sec Science Education	.41	.20	.34	-.07	.95	-
Years Teaching Experience	-	.22	.62	.99	.12	-
Days In-Service Training	-.44	.87	.66	-.25	.39	-
School Equipment						
Number of Labs	.87	-.12	1.0	.99	.95	.72
Use of Labs (Percent Time)	.56	.99	.06	.24	.29	.42
Lab Assistants	-	-	-	-	-	-
Like School						
School is not Enjoyable (r)	.73	.66	.65	.59	.54	.64
Enjoy Everything at School	.55	.39	.51	.43	.40	.39
Bored at School (r)	.72	.70	.71	.69	.69	.69
Do Not Like Many Subjects (r)	.76	.59	.45	.59	.57	.67
School Most Enjoyable	.27	.37	.35	.51	.48	.47
Want More Education	.46	.26	.28	.49	.26	.32
Dislike Schoolwork (r)	.68	.64	.66	.70	.74	.72
Experiments						
Students Do Experiments	.47	.94	.70	.71	.84	.34
Small Group Experiments	.99	.74	.82	.81	.70	.97
Teaching Style						
Teacher Reviews First	.34	.57	.59	.64	.54	.68
Teacher Starts with Explanation	.57	.60	.62	.60	.63	.54
Teacher Finished with Summary	.54	.64	.61	.58	.57	.56
Teacher Uses Students' Ideas	.21	.55	.53	.49	.21	.26
Students Set Own Problems; Teacher Helps	.58	.48	.55	.46	.67	.49
Teacher Sets Problems; Students Solve	.40	.49	.55	.47	.53	.05
Students Set Own Problems and Solve	.50	.42	.41	.37	.54	.26
Classroom Effort						
Teacher Asks You a Question	-	.51	.43	.50	.35	.48
Teacher Asks Someone Else a Question	-	.60	.49	.48	.53	.59
Pay Attention	-	.57	.59	.68	.75	.67
Make Up for Missed Lessons	-	.56	.66	.65	.59	.50
Deliberately Miss a Class	-	.57	.60	.57	.60	.56
Homework						
Hours Homework per Week in all Subjects	.92	.95	0.97	.99	.81	.96
Hours Homework in Science	.96	.95	0.96	.53	.87	.89
Time						
Hours Science instruction per week	.25	-.20	-.47	-0.1	.31	-
Amount of Labwork by Teacher	.88	.76	.52	-.62	.64	.60
Amount of Practical Work	.21	.25	.22	-.13	.85	-.75
Total Hours Teaching Instruction	.17	.83	.84	.81	-	.95
View of and Interest in Science						
Worth Spending Money on Science	.64	.38	.30	.36	.35	.33
Anxiety in Modern Society Due Science	.34	.16	.12	.23	.35	.19
People Understand Science Better Off	.33	.09	.13	.30	.23	.11
Inventions Improve Living Standards	.67	.11	.21	.23	.14	.15
Science is for Creative People	.29	.26	.28	.31	.22	.34
Science Causes World's Problems	.55	.19	.17	.27	.34	.34
Chemistry is an Enjoyable Subject	-	.68	.70	.72	.66	.74
School Chemistry Interesting	-	.73	.75	.68	.66	.71
Science Lessons Interesting	-	.54	.62	.30	.58	.36
Like Science	-	.50	.38	.66	.48	.53

Appendix I

TABLE I.8: Direct and total between construct effects and R-square (Population 3 - Chemistry)

Predicted/Predictor	Canada(Fr) D	Canada(Fr) T	Hong Kong (Form 6) D	Hong Kong (Form 6) T	Hong Kong (Form 7) D	Hong Kong (Form 7) T	Korea D	Korea T	Poland D	Poland T	Singapore D	Singapore T
Literacy of Home												
Socio-economic Status of Home	-	-	.09	.09	.06	.06	.47	.47	.46	.46	-.16	-.16
R^2	-		.009		.004		.222		.212		.025	
Like School												
Socio-economic Status of Home	-.13	-.13	-.05	-.04	-.11	-.10	-.07	-.02	-.09	-.05	-.04	-.03
Literacy of Home	-	-	.09	.09	.09	.09	.12	.12	.10	.10	-.07	-.07
Sex of Student	.18	.18	.07	.07	.12	.12	.16	.16	.08	.08	.07	.07
School Equipment	.06	.06	.02	.02	.03	.03	.01	.01	-.14	-.14	.07	.07
R^2	.033		.013		.017		.036		.036		.016	
School Science Equipment												
Location of School	.11	.11	-	-	-	-	-.33	-.33	-.05	-.05	-.33	-.33
Size of School	.72	.72	.18	.18	.28	.28	.36	.36	.40	.40	.50	.50
R^2	.520		.033		.078		.376		.175		.453	
Experiments												
Teacher Training and Experience	-.03	-.03	.12	.12	.03	.03	.04	.04	.02	.02	-	-
R^2	.001		.015		.001		.002		.000		-	
Homework												
Socio-economic Status of Home	.06	.06	.03	.05	.06	.06	.03	.04	.02	.07	.07	.06
Literacy of Home	-	-	.13	.13	.06	.06	.02	.02	.10	.10	.05	.05
Sex of Student	.54	.54	.56	.56	.76	.76	.15	.15	.17	.17	.01	.01
Teacher Training and Experience	-.04	-.04	.05	.05	.02	.02	-.03	-.03	-.06	-.06	-	-
R^2	.336		.363		.651		.027		.043		.006	
Classroom Effort												
Sex of Student	-	-	.03	.05	.04	.08	.06	.12	.16	.18	.07	.09
Like School	-	-	.30	.30	.33	.33	.37	.37	.29	.29	.29	.29
R^2	-		.093		.111		.144		.117		.092	
Time												
Teacher Training and Experience	-.24	-.24	.41	.41	.37	.37	.10	.10	.29	.29	-	-
School Equipment	.36	.36	.18	.18	.17	.17	.19	.19	-.13	-.13	-.57	-.57
R^2	.187		.208		.198		.059		.097		.320	
Views of and Interest in Science												
Socio-economic Status of Home	.07	.04	.05	.03	.07	.03	.04	.03	.12	.11	-.06	-.07
Sex of Student	-.10	-.06	-.08	-.05	-.08	-.03	-.13	-.07	-.07	-.02	-.17	-.14
Teacher Training and Experience	.00	.00	.04	.05	-.02	-.02	.03	.03	.05	.05	-	-
Experiments	.10	.10	.04	.04	.06	.06	.05	.05	.05	.05	.04	.04
Teaching Style	.01	.01	.16	.16	.22	.22	.05	.05	.21	.21	.06	.06
Like School	.22	.22	.36	.36	.35	.39	.30	.33	.24	.28	.31	.34
Classroom Effort	-	-	.12	.12	.14	.14	.09	.09	.14	.14	.13	.13
R^2	.063		.230		.262		.131		.187		.170	
Achievement												
Socio-economic Status of Home	-.05	-.05	-.02	-.01	.01	.02	.04	.16	.07	.10	.02	-.02
Literacy of Home	-	-	-.01	.01	.04	.04	.24	.26	.04	.05	.20	.20
Sex of Student	-.11	-.08	-.28	-.22	-.33	-.31	-.14	-.16	-.25	-.22	-.16	-.19
School Equipment	.01	-.00	-.07	-.02	.08	.12	.10	.12	-.02	.00	.36	.35
Teacher Training and Experience	.05	.06	.07	.21	.08	.16	.00	.01	-.05	-.03	-	-
Experiments	.10	.11	.16	.16	.01	.02	.00	.02	-.02	-.01	.08	.09
Teaching Style	-.18	-.17	-.12	-.10	-.08	-.06	-.15	-.13	-.08	-.06	-.10	-.08
Homework	.05	.05	.10	.10	.02	.02	-.01	-.01	.11	.11	.02	.02
Like School	.07	.11	-	.05	-	.06	-	.12	-	.04	-	.08
Classroom Effort	-	-	.03	.04	.07	.08	.07	.10	.05	.06	.01	.04
Time	-.07	-.07	.29	.29	.23	.23	.06	.06	.08	.08	.01	.01
Views of and Interest in Science	.15	.15	.11	.11	.09	.09	.29	.29	.08	.08	.23	.23
R^2	.095		.252		.233		.299		.103		.282	

D   direct effect
T   total effect

TABLE I.9: Summary of constructs and associated manifest variables (Population 3 - Physics)

	Canada (Fr)	Hong Kong (Form 6)	Hong Kong (Form 7)	Korea	Poland	Singapore
Socio-economic Status of Home						
Family Size	1.00	.81	.93	-.21	-.25	.99
Father's Education	-	.32	.13	.96	.92	-
Mother's Education	-	.31	.12	.87	.84	-
Father's Occupation	-	-	-	.63	-	-.02
Literacy of Home						
Word Knowledge Test	-	-	-	.83	-	.16
No. of Books in Home	-	.76	.86	.76	.99	.75
Use of Dictionary	-	.79	.66	.11	.40	.76
Teacher Training and Experience						
Years Post-Secondary Education	-.09	.69	.64	.62	-.20	-
Years Post-Sec Science Education	-.33	.16	.25	.81	.78	-
Years Teaching Experience	.78	.25	.66	-.45	-.60	-
Days In-Service Training	.69	.87	.67	.37	.37	-
School Equipment						
Number of Labs	.99	-.47	.99	1.00	.81	.65
Use of Labs (Percent Time)	.22	.96	.04	-.03	.55	.55
Like School						
School is not Enjoyable (r)	.73	.67	.64	.50	.42	.65
Enjoy Everything at School	.55	.37	.51	.37	.39	.40
Bored at School (r)	.71	.70	.71	.70	.76	.67
Do Not Like Many Subjects (r)	.68	.59	.46	.71	.49	.67
School Most Enjoyable	.28	.37	.35	.45	.56	.42
Want More Education	.55	.26	.30	.47	.20	.40
Dislike Schoolwork (r)	.69	.64	.66	.73	.79	.75
Experiments						
Students Do Experiments	.86	.94	.86	.18	.63	.91
Small Group Experiments	.82	.75	.65	-.48	.88	.59
Teaching Style						
Teacher Reviews First	.31	.58	.63	.49	.46	.50
Teacher Starts with Explanation	.26	.61	.63	.55	.58	.64
Teacher Finished with Summary	.46	.63	.62	.55	.67	.54
Teacher Uses Students' Ideas	.27	.52	.54	.65	.21	.20
Students Set Own Problems; Teacher Helps	.62	.49	.52	.50	.57	.42
Teacher Sets Problems; Students Solve	.62	.51	.54	.53	.58	.42
Students Set Own Problems and Solve	.57	.42	.44	.41	.50	.33
Classroom Effort						
Teacher Asks You a Question	-	.51	.43	.54	.24	.55
Teacher Asks Someone Else a Question	-	.58	.48	.57	.61	.61
Pay Attention	-	.60	.62	.69	.70	.73
Make Up for Missed Lessons	-	.55	.65	.68	.58	.61
Deliberately Miss a Class	-	.56	.59	.55	.59	.53
Homework						
Hours Homework per Week in all Subjects	.93	.95	.97	.95	.87	.97
Hours Homework in Science	.95	.95	.96	.77	.80	.82
Time						
Hours Science instruction per week	.77	-.18	-.43	.21	-.05	-
Amount of Labwork by Teacher	.72	.75	.53	-.08	.63	.77
Amount of Practical Work	-.20	.33	.31	.92	.87	.91
Total Hours Teaching Instruction	-.03	.86	.88	-.40	-	-.99
View of and Interest in Science						
Worth Spending Money on Science	.71	.43	.40	.34	.43	.36
Anxiety in Modern Society Due Science	.20	.18	.19	.15	.36	.08
People Understand Science Better Off	.52	.10	.19	.15	.05	.24
Inventions Improve Living Standards	.53	.12	.25	.32	.18	.17
Science is for Creative People	.45	.30	.32	.34	.22	.27
Science Causes World's Problems	.42	.21	.18	.07	.34	.19
Physics is an Enjoyable Subject	-	.66	.60	.75	.69	.77
School Physics Interesting	-	.38	.41	.64	.70	.74
Science Lessons Interesting	-	.54	.62	.24	.50	.32
Like Science	-	.55	.50	.70	.32	.56

TABLE I.10: Direct and total between construct effects and R-square (Population 3 - Physics)

Predicted/Predictor	Canada(Fr) D	Canada(Fr) T	Hong Kong (Form 6) D	Hong Kong (Form 6) T	Hong Kong (Form 7) D	Hong Kong (Form 7) T	Korea D	Korea T	Poland D	Poland T	Singapore D	Singapore T
Literacy of Home												
Socio-economic Status of Home	-	-	.10	.10	.06	.06	.48	.48	.53	.53	.01	.01
R^2			.010		.004		.230		.282		.000	
Like School												
Socio-economic Status of Home	-.12	-.12	-.05	-.04	-.10	-.10	-.06	.03	-.10	-.10	.05	.05
Literacy of Home	-	-	.09	.09	.09	.09	.19	.19	-.01	-.01	.09	.09
Sex of Student	.21	.21	.08	.08	.12	.12	.08	.08	.12	.12	.02	.02
School Equipment	.02	.02	.03	.03	.04	.04	-.01	-.01	-.03	-.03	-.03	-.03
R^2	.015		.014		.016				.031		.013	
School Science Equipment												
Location of School	.06	.06	-	-	-	-	-.43	-.43	.08	.08	-.54	-.54
Size of School	.74	.74	.14	.14	.28	.28	.29	.29	.50	.50	.36	.36
R^2	.548		.020		.076		.398		.211		.470	
Experiments												
Teacher Training and Experience	-.04	-.06	.13	.13	.08	.08	-.08	-.08	.06	.06	-	-
R^2	.004		.017		.006		.007		.004			
Homework												
Socio-economic Status of Home	.04	.04	.03	.05	.05	.06	.09	.11	.01	.07	.06	.06
Literacy of Home	-	-	.12	.12	.07	.07	.04	.04	.10	.10	.11	.11
Sex of Student	.54	.54	.56	.56	.77	.77	.13	.13	.25	.25	.05	.05
Teacher Training and Experience	-.09	-.09	.05	.05	.02	.02	.07	.07	.03	.03	-	-
R^2	.322		.359.		.653		.045		.068		.021	
Classroom Effort												
Sex of Student	-	-	.04	.06	.04	.08	.04	.07	.10	.12	.05	.06
Like School	-	-	.30	.30	.34	.34	.37	.37	.25	.25	.32	.32
R^2	-		.95		.115		.141		.080		.103	
Time												
Teacher Training and Experience	.36	.36	.41	.41	.39	.39	.40	.40	.20	.20		
School Equipment	.29	.29	.21	.21	.18	.18	-.09	-.09	-.24	-.24		
R^2	.256		.223		.223		.180		.098			
Views of and Interest in Science												
Socio-economic Status of Home	.04	.02	.10	.08	.16	.12	.11	.12	.08	.05	.00	.02
Sex of Student	-.17	-.12	-	.04	-.20	-.15	-.05	-.02	-.23	-.17	-.25	-.24
Teacher Training and Experience	.05	.05	.03	.04	.02	.02	.04	.04	.05	.05	-	-
Experiments	.08	.08	.06	.06	.02	.02	-.06	-.06	.03	.03	.02	.02
Teaching Style	.06	.06	.14	.14	.20	.20	.06	.06	.22	.22	.09	.09
Like School	-	-	.33	.37	.34	.34	.33	.39	.27	.31	.32	.37
Classroom Effort	-	-	.15	.15	.13	.13	.15	.15	.16	.16	.17	.17
R^2	.072		.220		.245		.197		.237		.234	
Science Achievement												
Socio-economic Status of Home	.02	.02	.01	.02	.03	.05	.02	.22	.07	.14	-.04	-.04
Literacy of Home	-	-	-	-	-	-	.33	.36	.09	.10	.01	.02
Sex of Student	-.14	-.13	-.14	-.08	-.26	-.14	-.08	-.09	-.21	-.20	-.27	-.29
School Equipment	-.07	-.07	-.13	-.06	.06	.11	.01	.01	.02	.06	.26	.22
Teacher Training and Experience	-.04	-.07	.07	.22	.02	.13	-.03	-.02	.11	.10	-	-
Experiments	.11	.12	.14	.15	-.02	-.01	-.04	-.06	.08	.09	.01	.01
Teaching Style	-.16	-.15	-.13	-.12	-.13	-.12	-.19	-.17	-.06	-.03	-.12	-.11
Homework	.05	.05	.12	.12	.17	.17	-.05	-.05	.15	.15	.02	.02
Like School	.06	.09	-	.04	-	.02	-	.14	-	.04	-	.08
Classroom Effort	-	-	.03	.04	-.02		.07	.11	-.01	.04	.07	.09
Time	-0.6	-.06	.30	.30	.28	.28	.02	.02	-.15	-.15	-.04	-.04
Views of and Interest in Science	.17	.17	.08	.08	.07	.07	.31	.31	.15	.15	.14	.14
R^2	.106		.203		.150		.352		.172		.166	

D direct effect
T total effect

Subject Index

	page(s)
Achievement	
adjusted scores (for Opportunity To Learn)	104-105
élite scores (Population 3)	70-71
increase across populations	73-76
levels of	93-94
patterns of	93-94
science subscores	57-58, 61-62, 199-200
science total scores	55-57, 60-61, 64-69, 195
scores	54-81
sex differences	75, 77
test scores for Populations 3E, 3X, 3N (Appendix D)	195
United States scores (Population 3)	69-70
Age of entry to schools	5
Age of student and achievement (Population 3)	71-73
Age of teacher	17, 19, 28-29, 38, 40-41
Attitudes (and descriptive scales)	
Beneficial Aspects of Science	83, 85-86, 201-216
Career Interest in Science	84, 88-89, 201-216
Classroom Effort	84, 89-90, 125-131, 201-216
Facility of Learning Science	83, 86, 201-216
Homework Effort	84, 89-90, 125-131, 201-216
Interest in Science	84, 87, 201-216
Like School	84, 86-87, 201-216
Practical Work	83, 85, 201-216
Student Directed Learning	84, 88, 201-216
Teacher Directed Learning	84, 89, 201-216
Conception of Learning Process	91-94
Country summaries	
Australia	133-134
Canada (Eng.)	134
Canada (Fr.)	134-135
China	135-136
England	136-137
Finland	137-138
Ghana	138-139
Hong Kong	139-141
Hungary	141-142
Israel	142-143
Italy	143-144
Japan	144-145
Korea	145-146
Netherlands	147
Nigeria	147-148
Norway	148
Papua New Guinea	148-149
Philippines	149-150
Poland	150-151
Singapore	151-153
Sweden	153-154
Thailand	154-155
United States	155-156
Zimbabwe	156

Curriculum
    course offerings   115-120
    decision levels   47, 97-100
    organization, content, implementation   158-159
    subject requirements   116-120
    system patterns of   119-124

Days schooling per year   14-15
Dissimilarity index-pattern   96-100
Descriptive scales (see Attitudes and descriptive scales)

Education budget   14-15
Equipment, school science (and laboratories)   20-21, 31-33, 39-44, 48,
                                                             125-132, 158
Experiments (and practical work)   21, 25-30, 36-41, 126
Experiments (construct)   125-132

Further education (expected)   34, 36-37

Grade-span (Population 2)   109-111
Grades taught (Population 2)   111-114
Gross National Product (GNP) per capita   14-16

Home
    Books in   16, 18
    Family background   45
    Literacy of (construct)   125-132, 157
    Socio-economic status of (construct)   125-132, 157
Homework
    all subjects   25-27, 35-37, 46, 162
    science subjects   25-27, 46, 125-132, 162
Homework (construct)   125-132

Influences on achievement   125-132
Instruction, total per year   20
Intra-class correlation (roh)   56, 60-61, 64-65
Item results (examples)   50-54

Laboratories (and equipment) (see Equipment)
Land area   14-15
Language used at home   35-37
Length of secondary schooling   72-73, 159-160

Measures of science achievement   49

Opportunity to Learn (OTL)
    and achievement   103-106
    direct measure   94-95, 100-106
    general   91-107, 121-123
    indirect measure   95-97

Pace of learning   163
Partial Least Squares (PLS) analyses   125-132, 221-234
Participants in study (Appendix A)   167-169
Percentage of an age group
    in school   5-6, 14-15, 44-45, 71-72, 160
    studying biology, chemistry, physics (Population 3)   6, 71-72, 160
Primary and secondary school science achievement relation   163
Practical work (see Experiments)
Practical Work (construct) (see Attitudes)
Pupil-teacher ratio   14-15

Reliability indices (KR-20) and alpha coefficients   197, 207-216

Subject Index

Sampling
    errors and bias   10
    marker variable   194
    procedures (*Appendix B, Part 2*)   7-8, 179-185
    response rates (*Appendices C1-C7*)   9, 187-193
    weighting   9-10
School learning experiences   92-93
Science subjects taught (Population 2)   115-120
Sex
    of school   20, 22, 30, 32-33, 42-44, 48
    of student (*see Achievement, sex differences*)   16, 18, 24, 26-27, 34, 36-37
    of teacher   17, 19, 25, 28-29, 32-33, 38, 40-41
Size
    of class   14, 15, 17-20, 24, 26-27, 35-37, 161-162
    of family   16, 18, 24, 26, 34, 36-37, 125-132
    of population   14-15
    of school   14-15, 20, 22, 39, 42-43, 47, 161
Structure of science teaching   21, 23-24
Subjects studied, number at terminal grade   6, 14-15, 71-72, 160-161

Target Population definitions (*Appendix B, Part 1*)   171-179
Teacher experience   17, 19, 38, 40-41, 125-132
Teacher salary   14-15, 20
Teacher training   17, 19, 28-30, 38, 40-41, 46, 125-132, 161
Teacher, time preparing and marking lessons   28-30, 38, 40-41, 47
Teaching style (construct)   125-132
Time
    allocated instructional hours   22-23, 32-33, 35-37, 42-43, 125-132
    construct   125-132
    homework (*see Homework*)
    hours per year science   25-26, 35-37, 39, 42-44, 125-132

View and Interest in Science (construct)   125-132, 223, 225